Current Topics in Microbiology and Immunology

107

Editors

Retroviruses 2

Edited by P.K. Vogt and H. Koprowski

With 26 Figures

Springer-Verlag
Berlin Heidelberg NewYork Tokyo 1983

Professor Dr. PETER K. VOGT
University of Southern California
School of Medicine
Department of Microbiology
2025 Zonal Avenue HMR 401
Los Angeles, CA 90033, USA

Professor Dr. HILARY KOPROWSKI
The Wistar Institute
36th Street at Spruce
Philadelphia, PA 19104, USA

ISBN-13: 978-3-642-69077-8 e-ISBN-13: 978-3-642-69075-4
DOI: 10.1007/978-3-642-69075-4

Softcover reprint of the hardcover 1st edition 1983

Library of Congress Catalog Card Number 15-12910

2123/3130-543210

Table of Contents

P. Besmer: Acute Transforming Feline Retroviruses. With 2 Figures 1

A. Tereba: Asymmetric Chromosomal Distribution of Endogenous Retrovirus Loci in Chickens and Mice. With 2 Figures 29

J.G. Krueger, E.A. Garber, and A.R. Goldberg: Subcellular Localization of $pp60^{src}$ in RSV-Transformed Cells. With 16 Figures 51

J.A. Cooper and T. Hunter: Regulation of Cell Growth and Transformation by Tyrosine-Specific Protein Kinases: The Search for Important Cellular Substrate Proteins. With 6 Figures . . 125

J.G. Foulkes: Phosphotyrosyl-Protein Phosphatases 163

Indexed in Current Contents

List of Contributors

BESMER, P., Memorial Sloan-Kettering Cancer Center, Sloan Kettering Division, Graduate School of Medical Sciences, Cornell University, 1275 York Avenue, New York, NY 10021, USA

COOPER, J.A., Molecular Biology and Virology Laboratory, The Salk Institute, P.O. Box 85800, San Diego, CA 92138, USA

FOULKES, J.G., Center for Cancer Research, Massachusetts Institute of Technology, Cambridge, MA 02139, USA

GARBER, E.A., The Rockefeller University, New York, NY 10021, USA

GOLDBERG, A.R., The Rockefeller University, New York, NY 10021, USA

HUNTER, T., Molecular Biology and Virology Laboratory, The Salk Institute, P.O. Box 85800, San Diego, CA 92138, USA

KRUEGER, J.G., The Rockefeller University, New York, NY 10021, USA

TEREBA, A., St. Jude Children's Research Hospital, 332 North Lauderdale, Memphis, TN 38101, USA

Acute Transforming Feline Retroviruses

PETER BESMER

1 Introduction . . . 1
2 Isolation and In Vivo Transformation Specificities of Feline ATVs . . . 3
3 Molecular Biology of Feline ATVs . . . 4
3.1 Feline Leukemia Virus (FeLV) . . . 5
3.2 v-*fes* ATVs: ST- and GA-FeSV . . . 7
3.3 v-*fms* ATVs: SM-FeSV . . . 10
3.4 v-*abl* ATVs: HZ2-FeSV . . . 11
3.5 v-*sis* ATVs: PI-FeSV . . . 14
3.6 GR-FeSV . . . 16
4 Origins of Feline ATVs . . . 16
4.1 Feline c-*oncs*: c-*fes* . . . 17
4.2 Mechanism of Transduction . . . 18
4.3 How Many Retroviral Oncogenes Are There? . . . 20
References . . . 21

1 Introduction

The study of retroviruses associated with the domestic cat (*Felis catus*) has been of interest mainly because of the infective nature of these viruses in the domestic cat population and their association with neoplasms of an outbred mammalian species. The feline retroviruses can be divided into three groups: (a) the feline leukemia viruses (FeLVs), (b) the endogenous, xenotropic feline retroviruses, and (c) the acute transforming feline retroviruses (ATVs).

Feline leukemia viruses have been divided into subgroups according to their host range and interference properties (SARMA and LOG 1971, 1973; JARRETT 1980). All of them replicate in feline cells and they therefore are ecotropic viruses. FeLVs are horizontally transmitted in the cat population and are associated with a wide spectrum of neoplastic and degenerative blastopenic diseases: lymphosarcoma, erythremic myelosis, erythroleukemia, megakaryocytic leukemia, granulocytic leukemia, fibrosarcoma, erythroblastosis, pancytopenia, myeloblastopenia, glomerulonephritis and thymic atrophy (HARDY et al. 1973; JARRETT et al. 1973a; HARDY 1980). The FeLVs

Memorial Sloan-Kettering Cancer Center, Sloan-Kettering Division, Graduate School of Medical Sciences, Cornell University, 1275 York Avenue, New York, NY 10021, USA

Current Topics in Microbiology and Immunology. Vol. 107

are chronic leukemia viruses, i.e., they induce neoplasms with a long latency period and do not transform fibroblasts in culture (RICKARD et al. 1969). Similar to other retroviruses, the integrated provirus of FeLV contains LTR sequences at both the 5′ and the 3′ end and the three structural genes gag, pol, and env, all of which are needed for the replication of the virus. DNA sequences homologous to FeLV are present in the genome of uninfected domestic cats (BENVENISTE et al. 1975). These endogenous FeLV-related sequences, about seven to nine copies per haploid genome, do not represent complete FeLV genomes, suggesting that FeLV is maintained in the cat population by exogenous infection. Nucleic acid sequence homology exists between FeLV and rodent retroviruses (BENVENISTE et al. 1975). It is, therefore, believed that FeLVs derive from an ancestral rodent retrovirus by cross-species infection.

The *xenotropic endogenous feline retroviruses,* which include the RD114 and the CCC viruses (MCALLISTER et al. 1972), constitute a separate class of feline retroviruses. They are distinct from FeLV by nucleic acid and by serological criteria as well as by host range and interference properties (EAST et al. 1973; TODARO et al. 1974; SHERR and TODARO 1974; HENDERSON et al. 1974). Sequences homologous to the CCC/RD114 virus are found in the genome of domestic cats (TODARO et al. 1974; BALUDA and ROY-BURMAN 1973). In contrast to the FeLV proviruses contained in the genome of uninfected cats, the endogenous CCC/RD114-related sequences represent complete viral genomes (LIVINGSTON and TODARO 1973). The viruses of the CCC/RD114 group are not oncogenic. Interestingly, the xenotropic feline retroviruses are related to the Baboon endogenous virus.

Acute transforming feline retroviruses (feline ATVs) were isolated from FeLV-associated feline multicentric fibrosarcomas. They induce neoplasms with a short latency period in animals and transform tissue culture cells in vitro. All feline ATVs are replication defective and need a helper virus for their propagation. Their genomes contain FeLV-related sequences and transformation-specific sequences (v-*onc* sequences) (FRANKEL et al. 1979). The v-*onc* sequences in these viruses are homologous with single-copy cellular genes (c-*oncs*), and ATVs therefore are thought to have arisen by recombination of FeLV with cat c-*onc* sequences (FRANKEL et al. 1979; SHERR et al. 1979). The oncogenic properties of ATVs are believed to be determined by the v-*onc*c sequences in these viruses. Until now, four different v-*onc* genes have been found in feline ATVs. One of the feline v-*onc* genes, v-*fms,* is unique. The other three, v-*fes,* v-*sis,* and v-*abl,* have also been identified in ATVs isolated from other species (WEISS et al. 1982).

The main purpose of this paper is to review the molecular biology of feline ATVs. Specifically, I will discuss the history of their isolation, their in vivo and in vitro transformation specificity, their genome structure, viral RNA and protein expression, and the viral oncogenes and their origins. Where appropriate I will compare feline ATVs with acute transforming viruses from other species with homologous v-*onc* genes. Since all feline ATVs derive by recombination from FeLV and cat cellular sequences, I will also briefly discuss the molecular biology of FeLV.

2 Isolation and In Vivo Transformation Specificities of Feline ATVs

Fibrosarcomas account for 6%–12% of all cat neoplasms (HARDY 1981). They are found mostly in older cats (mean age 10 years) and tend to be slow-growing solitary tumors. No ATVs have been isolated from old cats with solitary fibrosarcomas. Less frequently, multicentric fibrosarcomas are found in younger cats (mean age 3 years). These multicentric fibrosarcomas are often associated with FeLV infection. Feline acute transforming viruses have been obtained from the FeLV-associated multicentric fibrosarcomas of younger cats. Historically, sarcoma viruses were identified and isolated by induction of neoplasms in fetal or neonatal kittens with cell-free tumor extracts (SNYDER and THEILEN 1969). More recently, transforming agents have also been identified by focus induction in tissue culture cells with cell-free tumor extracts (HARDY et al. 1982; BESMER et al. 1983b).

The first isolation of a feline sarcoma virus (FeSV) was reported by Snyder and Theilen in 1969 (SNYDER and THEILEN 1969). Subcutaneous injection of cell-free tumor extracts from a cat with multiple subcutaneous fibrosarcoma and multiple metastasis was shown to induce fibrosarcomas in newborn kittens. This first acute transforming feline retrovirus, the Snyder-Theilen feline sarcoma virus, was designated ST-FeSV. The second feline ATV, the Gardner-Arnstein FeSV (GA-FeSV), was isolated from a Siamese cat with recurrent multiple subcutaneous fibrosarcoma (GARDNER et al. 1970). Cell-free tumor extracts again were found to induce fibrosarcoma in fetal and newborn kittens. When injected subcutaneously, the GA-FeSV induced melanomas and fibrosarcoma at the site of injection, while upon intraocular injection only melanomas were obtained (MCCULLOUGH et al. 1972; NIEDERKORN et al. 1980). The GA-FeSV therefore transforms cells of mesodermal and ectodermal origin. Experimental infection of species other than the cat with ST- and GA-FeSV demonstrated that fibrosarcomas can be induced in a wide range of different mammals, including guinea pigs, rabbits, dogs, goats, sheep, and a number of primates (reviewed in

Table 1. Characteristics of acute transforming feline retroviruses

Strain designation		Types of neoplasms induced in kittens	v-*onc*	Protein product of v-*onc*	Protein kinase activity
Snyder-Theilen	FeSV (ST-FeSV)	Fibrosarcoma	*fes*	P85 *gag-fes*	+
Gardner-Arnstein	FeSV (GA-FeSV)	Fibrosarcoma and melanoma	*fes*	P95 *gag-fes*	+
Hardy-Zuckerman 1	FeSV (HZ1-FeSV)	Fibrosarcoma	*fes*	P96 *gag-fes*	+
McDonough	FeSV (SM-FeSV)	Fibrosarcoma	*fms*	gp170 *gag-fms*	–
Parodi-Irgens	FeSV (PI-FeSV)	Fibrosarcoma	*sis*	P76 *gag-sis*	–
Gardner-Rasheed	FeSV (GR-FeSV)	Fibrosarcoma	?	P70 *gag-onc* ?	+
Hardy-Zuckerman 2	FeSV (HZ2-FeSV)	Fibrosarcoma	*abl*	P96 *gag-abl*	+

HARDY 1981). Transformation by ST- and GA-FeSV thus is not limited to the feline species. Murine and avian sarcoma viruses similarly have been shown to induce fibrosarcomas in other species. Pseudotypes of GA-FeSV with Moloney-MuLV and with amphotropic MuLV have been used to infect neonatal NFS mice, but no tumors were detected in the infected animals (EVEN et al. 1983). This is surprising since the same virus preparations easily induce transformation in murine cells in vitro.

The SM-FeSV was isolated in 1971 by MCDONOUGH and collaborators (MCDONOUGH et al. 1971). This virus was obtained from a $1^1/_2$ years old domestic short-hair cat with recurrent multiple subcutaneous fibrosarcoma. Another ATV strain, the Parodi-Irgens FeSV (PI-FeSV), was isolated in France (IRGENS et al. 1973). The PI-FeSV derived from a $1^1/_2$ year old cat with multiple subcutaneous fibrosarcoma with metastatic lesions in the lung and the peritoneum. This virus too induced sarcomas in newborn kittens. The Gardner-Rasheed FeSV (GR-FeSV) was isolated in 1982 (RASHEED et al. 1982). The fibrosarcoma from which the GR-FeSV was isolated contained a rhabdomyosarcomatous component, and upon intramuscular injection into kittens, neoplasms of the same type were obtained with metastasis to skeletal, visceral, and central nervous system tissues. In addition Snyder reported three transmissible fibrosarcomas: from a 14-month-old cat with multiple subcutaneous fibrosarcoma, from a 7-year-old cat with a solitary fibrosarcoma in the thorax with metastasis in the lung and the pleural cavity, and from a second 7-year-old cat with solitary fibrosarcoma in the deep fascia of the neck. Unfortunately the viruses isolated from these tumors have been lost (SNYDER 1971).

Two ATVs have recently been isolated in Hardy's laboratory. In contrast to earlier isolations of feline ATVs, the Hardy-Zuckerman strains of FeSV, HZ1-FeSV and HZ2-FeSV, were isolated by tissue culture procedures (SNYDER et al., in preparation; BESMER et al. 1983b). Both the HZ1- and the HZ2-FeSV derive from multicentric fibrosarcomas of pet cats. Upon injection of cell free tumor extracts into kittens, both viruses the HZ2-FeSV again induced fibrosarcomas.

All of the known feline ATVs induce tumors in vivo of the same type from which the virus originally was isolated. Only one, the GA-FeSV, was found to cause different types of neoplasms, namely fibrosarcomas and melanomas. No systematic studies have, however, been carried out, e.g., using different routes of inoculation in order to identify in vivo transformation targets other than those of the fibroblast cell lineage.

3 Molecular Biology of Feline ATVs

All the original isolates of feline ATVs were mixtures of ATV and FeLV. Very often the FeLV helper virus is present in large excess (100–1000-fold) (SARMA et al. 1971a; BESMER et al. 1983a). This situation has masked the replication-defective nature of the sarcoma viruses. The isolation of virus nonproducer cells, i.e., cells that are infected by the ATV in the absence of helper virus, was therefore an important step in the analysis of the genomes and the transcriptional and translational products of these viruses (CHAN et al. 1974; HENDERSON et al. 1974; PORZIG et al. 1979). Transformed virus nonproducer cells can be obtained by low multiplicity infection of cells and isolation of foci of morphologically transformed cells. The most common cell lines used for the generation of virus nonproducer cells are the mink lung cell line (CCL 64), FRE and NRK rat cells, and NIH/3T3 mouse cells. For the propagation of the ATVs these can be rescued from the virus nonproducer cells by superinfection with replication-competent mammalian retroviruses such as the amphotropic MuLV, FeLV, or Moloney-MuLV. The amphotropic murine leukemic virus is particularly versatile for this purpose because of its broad host range (RASHEED et al. 1976) and because phenotypic mixing between FeLV and its derivative ATVs with murine retroviruses is a very efficient process (EVEN et al. 1983).

Of great importance in the study of feline ATVs was the demonstration of the recombinational origin of ATVs from FeLV and cat cellular sequences (FRANKEL et al. 1979). The techniques employed in these early studies were analogous to those used for the characterization of ATVs in other species at the time and involved the preparation of ATV-specific cDNA hybridization probes (STEHELIN et al. 1976). Since then the molecular structure and the genetic content of the genomes of the ST-, GA-, and SM-FeSV have been studied using recombinant DNA techniques. More recently, new isolates of feline ATVs have been obtained. Their characterization indicates that FeLV transduces a surprising variety of cellular oncogens in the pet cat population and that retroviruses of different species are able to transduce homologous cellular oncogenes.

3.1 Feline Leukemia Virus (FeLV)

Feline leukemia viruses have been divided into three subgroups, A, B, and C, according to their host range and interference properties (SARMA and LOG 1971, 1973; JARRETT et al. 1973b; JARRET (1980a, b). They are ecotropic retroviruses, that is, all of them replicate on cat cells. FeLV-A has the narrowest host range, growing preferentially on cat cells; FeLV-B and FeLV-C on the other hand replicate also on cells from other species such as mink, dog, and human (SARMA et al. 1975). All naturally occurring FeLV isolates contain FeLV-A, so that isolates of FeLV-B and FeLV-C are always contained in mixtures with FeLV-A (JARRETT 1980a, b).

By liquid nucleic acid hybridization techniques, the different subgroup FeLVs are highly related (ROBBINS et al. 1979). Using a more sensitive method, T1 oligonucleotide RNA fingerprinting, Rosenberg and Hazeltine found a remarkable amount of sequence identity between different FeLV strains (ROSENBERG et al. 1981). No oligonucleotide maps, however, were made from any of the analyzed FeLVs in order to determine whether differences observed in the various strains map to a particular region of the FeLV

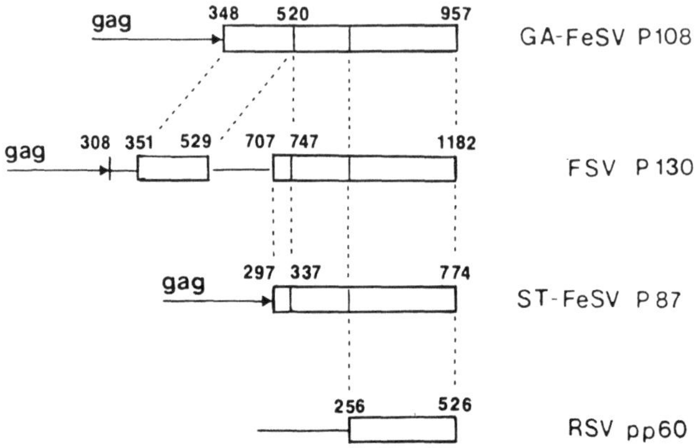

Fig. 1. Homologous amino acid domains in ST-FeSV, GA-FeSV, Fujinami sarcoma virus (FSV), and Rous sarcoma virus (RSV) transforming proteins. The *gag-onc* junctions in ST-FeSV, GA-FeSV, and FSV polyproteins are marked by *arrowheads*. Homologous domains are shown as *open rectangles* and are connected by *dotted lines*. (HAMPE et al. 1982)

been published (HAMPE et al. 1982; SHIBUYA and HANAFUSA 1982). Their comparison revealed interesting features of the related transforming proteins (Fig. 1). At the N-terminus the *gag-fes* proteins of GA- and ST-FeSV contain 348 and 297 amino acids each of the *gag* sequences p15, p12, and p30. The GA strain contains 151 amino acids of p30 sequences and the ST strain 100 p30 amino acids. v-*fes* of the ST and the GA strains encode 477 and 609 amino acids respectively. The C-terminal 437 amino acids of the ST and the GA *gag-fes* proteins are almost completely homologous. The 3′ ends of both the ST and the GA v-*fes* sequences terminate 51 nucleotides beyond the polyprotein coding sequences. Interestingly, the 3′ v-*fes* sequences are joined to identical FeLV sequences in both the ST- and the GA-FeSV. The FSV genome encodes a 130000-dalton *gag-fps* polyprotein. It contains 308 amino acids derived from the ALV *gag* genes p15, p10, and p27 and 871 amino acids derived from the *fps* sequences. Between the FSV, Ga-FeSV, and ST-FeSV *gag-fps* and *gag-fes* proteins there is an overall homology of approximately 70%. At the 5′ end the v-*fes* sequences of the GA- and the ST-FeSV strain differ. These strain-specific v-*fes* sequences of both the ST- and the GA-FeSV are represented in the FSV v-*fps* gene (Fig. 1). The ST-specific sequences are found in the middle of v-*fps* and those of the GA strain further upstream near the 5′ end of the FSV v-*fps*. With respect to v-*fps*, the GA-FeSV v-*fes* sequences therefore appear to contain an internal deletion (Fig. 1). It seems likely that the GA-FeSV derived by deletion from a progenitor similar to FSV upon passage of the virus in vitro or in vivo. This sequence comparison also suggests that the N-terminal 440 amino acids of v-*fps* may not be essential for the neoplastic properties of FSV. In the carboxyterminal domain the GA-FeSV, the ST-FeSV, and the FSV v-*fes* and v-*fps* sequences show extensive homology (85%). The extensive structural and functional relationship between v-*fps*

3.2 v-*fes* ATVs: ST- and GA-FeSV

The GA- and the ST-FeSV are replication-defective ATVs. Both of them transform fibroblastic and epithelioid cells in culture, and transformed virus nonproducer cell lines were derived in mink CCL64 and in rat NRK cells (HENDERSON et al. 1974; PORZIG et al. 1979). An initial understanding of the genetic composition of the genomes of the ST- and the GA-FeSV was obtained from liquid hybridization experiments such as those employed in the characterization of the v-*src* sequences of Rous sarcoma virus by STEHELIN and co-workers (STEHELIN et al. 1976; FRANKEL et al. 1979). A cDNA was synthesized from viral RNA of FeSV (FeLV) pseudotype virus. cDNA sequences specific to the FeSV genome were prepared by sequential adsorption with FeLV RNA and RNA isolated from FeSV nonproducer cells and hydroxyapatite chromatography. The genomes of the ST- and the GA-FeSV were found to be composed of sequences related to FeLV and sequences specific to both the ST- and the GA-FeSV (*fes* sequences). Sequences homologous to the *fes* cDNA were found in normal cat DNA at an abundance corresponding to a single copy gene. These experiments thus indicated the recombinational origins of the ST- and the GA-FeSV.

In 1980 Shibuya and collaborators found that the *fps* sequences of the Fujinami avian sarcoma virus are homologous with the *fes* sequences of the ST- and the GA-FeSV (SHIBUYA et al. 1980). This was the first demonstration that retroviruses from different species are able to transduce homologous cellular genes (c-*oncs*). The *fes* oncogene is the most prevalent oncogene found in acute transforming retroviruses. At present three feline ATVs and five avian ATVs are known to contain this oncogene.

More insight into the organization of the ST-FeSV genome was obtained from restriction mapping of unintegrated proviral DNA isolated from cells after infection and of molecularly cloned FeSV DNA genomes (SHERR et al. 1979, 1980a, b; FEDELE et al. 1981). The ST- and the GA-FeSV have a very similar genome organization; they have RNA genomes of 4.5 and 6.2 kb respectively. At the 5′ end both of them contain 1.3–1.4 kb of sequences related with FeLV and at the 3′ end the ST-FeSV contains 1.3–1.4 kb and the GA-FeSV 2.8 kb of FeLV-related sequences. The *fes* sequences in the middle consist of segments of 1.6 kb and 2.0 kb respectively. The genome structure in both the ST-FeSV and the GA-FeSV therefore appears to be 5′-*Δgag-fes-Δenv* 3′. This then predicts that the 85000- and 95000-dalton *gag*-polyproteins found in ST- and GA-FeSV-infected cells are *gag-fes* fusion proteins (STEPHENSON et al. 1977; SHERR et al. 1978; SNYDER et al. 1978). The genome organization of the ST- and the GA-FeSV is reminiscent of that of Abelson murine leukemia virus, the Fujinami, PRCII, and Y73 strains of avian sarcoma virus, and the avian acute transforming viruses, MC29, MH2, CMII, and E26, where helper virus *gag* sequences are fused to v-*onc* sequences to give rise to *gag*-v-*onc* fusion protein products (WEISS et al. 1982).

The nucleic acid sequences of the *gag-fes* genes of the ST- and the GA-FeSV and that of the Fujinami avian sarcoma virus (FSV) have recently for initiation and maintenance of transformation (BARBACID 1981; EVEN et al. 1983) Both the in vitro mutagenesis studies and the investigation of the td mutants have identified the *gag-fes* protein as the transforming protein of this virus.

3.3 v-*fms* ATVs: SM-FeSV

Like all other feline sarcoma viruses, the SM-FeSV is replication defective, and the isolation of transformed virus nonproducer cells made possible the characterization of the viral genome and it's gene products (PORZIG et al. 1979). Initial liquid hybridization experiments showed that the SM-FeSV genome is composed of FeLV-related and FeSV-specific nucleic acid sequences. The SM-FeSV specific sequences were designated v-*fms* and were shown to be homologous with a single-copy cat cellular gene (FRANKEL et al. 1979). These early studies indicated the recombinational origin of the SM-FeSV form FeLV and cat cellular sequences. The following structural analysis of a molecularly cloned integrated provirus provided more information about the organization of the SM-FeSV genome (DONNER et al. 1982). The virus has a 7.8-kb RNA genome. At the 5′ end it contains 2.1 kb of FeLV related sequences. The v-*fms* sequences are located on a 3.1-kb contiguous segment in the middle of the genome. Similar to GA-FeSV and ST-FeSV v-*fms* is joined to FeLV sequences 5′ of the *env* gene of FeLV. The gene order in the SM-FeSV thus was found to be *5′-Δgag-fms-env3′*. By nucleic acid hybridization v-*fms* is not related to v-*fes*, v-*mos*, v-*abl*, v-*kis*, v-*has*, v-*sis*, and v-*src*. It is believed to be a unique oncogene. The sequence relationship between v-*fms* and the avian oncogenes v-*myc*, v-*erb*, v-*myb*, and v-*yes*, however, has not been reported yet.

Cells infected by the SM-FeSV synthesize full-length genome RNA and a subgenomic *env* mRNA. The following protein products which are specified by the SM-FeSV genome have been detected in infect cells: 170000-dalton gag polyprotein, a gp140, a gp120, and a *gag* p60 protein (BARBACID et al. 1980a; RUSCETTI et al. 1980). The gene order 5′ *gag-fms* predicts that the 170000-kilodalton *gag* polyprotein is a *gag-fms* fusion protein. Hyperimmune antisera which recognize v-*fms* sequences in the 170000-dalton *gag-fms* protein were reported some time ago (BARBACID et al. 1980; RUSCETTI et al. 1980; VAN DE VEN et al. 1980b). More recently monoclonal antibodies to epitopes of the v-*fms* sequences have been prepared by two groups (VERONESE et al. 1982; ANDERSON et al. 1982). The v-*fms* specific monoclonal antibodies reacted with three different proteins: the gp170 *gag-fms* polyprotein, a gp140 *fms*, and a gp-120 *fms* protein. Steady state analysis of these proteins in infected cells showed that the gp140 and gp120 species are the predominant products. In contrast to other transforming proteins, the v-*fms* protein products are extensively glycosylated. Upon tunicamycin treatment of infected cells a 155000-dalton unglycosylated *fms* containing protein was found, indicating that the primary translation product is a 155000-dalton *gag-fms* protein which is glycosylated to yield the gp170 *gag-fms* protein.

and v-*fes* indicates that the two viral oncogens derive from homologous cellular genes of cats and birds. Of considerable interest was the finding of an amino acid homology of about 40% between the C-terminal domains of the *gag-fes* protein (about 40%) and those of the transforming protein pp60*src* of Rous sarcoma virus and the transforming protein of the avian sarcoma virus Y73 (Fig. 1). The relationship between v-*fps*, v-*src*, and v-*yes*, all of them of avian origin, suggests that these genes have evolved from a common ancestral gene.

Cells infected by the ST- and the GA-FeSV synthesize only full-length genome RNA. No subgenomic RNAs have been reported (SHERR, personal communication). In agreement with the nucleotide sequence of the two viruses, *gag-fes* polyproteins of 85000 and 95000 daltons were identified in infected cells (BARBACID et al. 1980a; RUSCETTI et al. 1980). These polyproteins contain antigenic determinants of the FeLV *gag* proteins p15, p12, and p30. Two minor glycosylated forms of the *gag-fes* polyprotein have also been observed with electrophoretic mobilities of 80000 daltons and 100000 daltons (SHERR et al. 1980b). Similar to the transforming proteins of Rous sarcoma virus and Abelson murine leukemia virus, the *gag-fes* proteins exhibit a tyrosine-specific protein kinase activity in immunoprecipitates. Substrates for the in vitro phosphorylation include the *gag-fes* protein as well as casein. Phosphorylation of immunoglobulin heavy chain has been observed by some investigators but not by others (BARBACID et al. 1980b; VAN DE VEN et al. 1980a; SNYDER 1982). Very likely as a consequence of the viral protein kinase activity, cells transformed by the ST- and the GA-FeSV have elevated levels of proteins containing phosphotyrosine. The two viral *gag-fes* proteins have a single site for tyrosine phosphorylation (BLOMBERG et al. 1981). In analogy with Rous sarcoma virus, this single phosporylation site has been predicted to be in the C-terminal domain of the *gag-fes* proteins. This predicted position is in agreement with recent sequence analysis of phosphotyrosine peptides (PATCHINSKY et al. 1982).

In order to investigate the role of the *gag-fes* protein in the initiation and maintenance of cell transformation, transformation-defective mutants (td mutants) of ST-FeSV were isolated by two groups (DONNER et al. 1980; REYNOLDS et al. 1981b). Cells which harbor transmissible td FeSV variants exhibit a nontransformed phenotype: they do not grow in semisolid medium, they bind high levels of epidermal growth factor, and they express the *gag-fes* polyprotein. Cells containing the mutant virus, however, do not have elevated levels of proteins containing phosphotyrosine, and the *gag-fes* protein is inactive in the in vitro phosphorylation assay (BARBACID et al., 1981; REYNOLDS et al. 1981b). These observations suggest that the protein kinase activity which is associated with the *gag-fes* protein is required for the transformation induced by the virus. The sequences of the GA-FeSV genome which are required for transformation have recently also been defined by in vitro mutagenesis of the molecularly cloned GA-FeSV provirus. The 5′ LTR sequences which contain the signals for initiation of transcription and enhancer elements which regulate the efficiency of transcription as well as the coding region for the entire *gag-fes* protein were found to be necessary for initiation and maintenance of transformation (Barbacid 1981; Even et al. 1983) Both the in vitro mutagenesis studies and the investigation of the td mutants have identified the *gag-fes* protein as the transforming protein of this virus.

3.3 v-*fms* ATVs: SM-FeSV

Like all other feline sarcoma viruses, the SM-FeSV is replication defective, and the isolation of transformed virus nonproducer cells made possible the characterization of the viral genome and it's gene products (Porzig et al. 1979). Initial liquid hybridization experiments showed that the SM-FeSV genome is composed of FeLV-related and FeSV-specific nucleic acid sequences. The SM-FeSV specific sequences were designated v-*fms* and were shown to be homologous with a single-copy cat cellular gene (Frankel et al. 1979). These early studies indicated the recombinational origin of the SM-FeSV form FeLV and cat cellular sequences. The following structural analysis of a molecularly cloned integrated provirus provided more information about the organization of the SM-FeSV genome (Donner et al. 1982). The virus has a 7.8-kb RNA genome. At the 5′ end it contains 2.1 kb of FeLV related sequences. The v-*fms* sequences are located on a 3.1-kb contiguous segment in the middle of the genome. Similar to GA-FeSV and ST-FeSV v-*fms* is joined to FeLV sequences 5′ of the *env* gene of FeLV. The gene order in the SM-FeSV thus was found to be *5′-Δgag-fms-env3′*. By nucleic acid hybridization v-*fms* is not related to v-*fes*, v-*mos*, v-*abl*, v-*kis*, v-*has*, v-*sis*, and v-*src*. It is believed to be a unique oncogene. The sequence relationship between v-*fms* and the avian oncogenes v-*myc*, v-*erb*, v-*myb*, and v-*yes*, however, has not been reported yet.

Cells infected by the SM-FeSV synthesize full-length genome RNA and a subgenomic *env* mRNA. The following protein products which are specified by the SM-FeSV genome have been detected in infect cells: 170000-dalton gag polyprotein, a gp140, a gp120, and a *gag* p60 protein (Barbacid et al. 1980a; Ruscetti et al. 1980). The gene order 5′ *gag-fms* predicts that the 170000-kilodalton *gag* polyprotein is a *gag-fms* fusion protein. Hyperimmune antisera which recognize v-*fms* sequences in the 170000-dalton *gag-fms* protein were reported some time ago (Barbacid et al. 1980; Ruscetti et al. 1980; Van de Ven et al. 1980b). More recently monoclonal antibodies to epitopes of the v-*fms* sequences have been prepared by two groups (Veronese et al. 1982; Anderson et al. 1982). The v-*fms* specific monoclonal antibodies reacted with three different proteins: the gp170 *gag-fms* polyprotein, a gp140 *fms*, and a gp-120 *fms* protein. Steady state analysis of these proteins in infected cells showed that the gp140 and gp120 species are the predominant products. In contrast to other transforming proteins, the v-*fms* protein products are extensively glycosylated. Upon tunicamycin treatment of infected cells a 155000-dalton unglycosylated *fms* containing protein was found, indicating that the primary translation product is a 155000-dalton *gag-fms* protein which is glycosylated to yield the gp170 *gag-fms* protein.

The gp170-protein possibly then is cleaved to yield the gp120 and gp140 *fms* proteins and the p60 *gag* protein. The v-*fms* containing protein products localize to the cytoplasm of infected cells and are quantitatively associated with sedimentable organelles. No significant tyrosine-specific protein kinase activity has been found to be associated with the *gag-fms* protein nor is this protein phosphorylated in vivo (REYNOLDS et al. 1981a). Similarly, no elevated levels of phosphotyrosine-containing proteins were detected in infected cells. When coprecipitated with the ST- or the GA-FeSV *gag-fes* proteins, a low level of gp170 *gag-fms* phosphorylation was observed and this phosphorylation was shown to be specific for tyrosine (VERONESE et al. 1982). Thus although the gp170 *gag-fms* protein lacks detectable protein kinase activity, it contains acceptor sites for tyrosine phosphorylation. From this evidence it seems probable that v-*fms* is not a member of the tyrosine kinase gene family. A comparison of the nucleotide sequence of v-*fms* with v-*fes*, v-*src*, v-*yes*, v-*abl*, and v-*mos* will, however, be necessary to confirm this prediction. No transformation-defective mutants have been isolated, nor have in vitro mutagenesis studies been reported to support the assumption that the *gag-fms* polyprotein and/or its cleavage products are the transforming protein of the SM-FeSV.

3.4 v-*abl* ATVs: HZ2-FeSV

The Hardy-Zuckerman 2 strain of FeSV (HZ2-FeSV), like all other FeSVs, derives from a cat with multicentric fibrosarcoma (BESMER et al. 1983b). The genome and the gene products of this replication-defective acute transforming retrovirus were analyzed in transformed virus nonproducer mink cells. The HZ2-FeSV provirus in the cellular DNA of mink virus nonproducer cells and the v-*onc* sequences contained in this virus were investigated by Southern blot analysis. It was found that restriction fragments which contained HZ2-FeSV sequences detected by an FeLV *rep* hybridization probe also hybridized with a v-*abl* hybridization probe, indicating that FeLV and v-*abl* sequences are linked in the HZ2-FeSV genome (BESMER et al. 1983b). It therefore appears that the v-*onc* sequences in the HZ2-FeSV are homologous with the v-*abl* sequences of the Abelson murine leukemia virus. The integrated provirus of the HZ2-FeSV has recently been cloned and analyzed in the author's laboratory. At the 5′ end it is homologous with 5′ FeLV sequences, including coding sequences of the FeLV *gag* proteins p15, p12, and p30, and at the 3′ end the viral genome contains FeLV *env* sequences (BERGOLD and BESMER, unpublished data). The size and the genetic composition of the cat v-*abl* insert is not precisely known yet. At the 5′ end v-*abl* is joined to FeLV *gag* sequences and at the 3′ end to FeLV sequences 5′ of the *gag-pol* junction. The gene order in the HZ2-FeSV therefore is 5′-*Δgag-abl-env*3′. Virus nonproducer cells synthesize 6.3-kb genome RNA and a 2.3-kb subgenomic RNA. The 6.3-kb RNA contains 5′ *gag*, *abl*, and *env* sequences but lacks *pol* sequences. The subgenomic RNA, on the other hand, contains only FeLV *env* sequences. This suggests that the v-*abl* gene product in this virus is expressed from the 6.3-kb RNA and not from the subgenomic RNA. A 98000-dalton *gag* polyprotein is synthesized in the mink HZ2-FeSV nonproducer cells. It contains determinants of the FeLV *gag* proteins p15, p12, and p30. The genome structure discussed above predicts that this *gag* polyprotein is a *gag-abl* fusion protein. When assayed for in vitro kinase activity in immunoprecipitates containing the 98000-dalton protein, the 98000-dalton protein as well as immunoglobulin heavy chain was phosphorylated and phosphorylation was shown to be specific for tyrosine residues (LEDERMAN and SNYDER, unpublished data). No FeLV *env* gene products were detected in HZ2-FeSV infected cells, indicating that the *env* gene in this virus is defective.

Transformed mink CCL64 virus nonproducer cells were obtained by infection of mink cells with the HZ2-FeSV complex, which contains subgroup A and subgroup B FeLV helper virus. Transformed mink cells are round, refractile, and very loosely attached to the tissue culture plates. The HZ2-FeSV can be rescued with amphotropic MuLV. Transformation indistinguishable from Abelson-MuLV is obtained after infection of NIH/3T3, Balb/3T3, and FRE-3a rat cells with HZ2-FeSV (amph-MuLV) pseudotype virus. No transformation is observed upon infection of FLF-3 and Fea cat fibroblasts (ZUCKERMAN and HARDY, unpublished data). Upon injection of three 6-week-old kittens with cell-free tumor extract, one kitten developed multicentric fibrosarcoma. No in vivo experiments have, however, been done with the biologically cloned HZ2-FeSV.

Abelson murine leukemia virus was generated by recombination of Moloney murine leukemia virus and c-*abl* sequences of a Balb/c mouse (ABELSON and RABSTEIN 1970a, b; GOFF et al. 1980; ROSENBERG 1982; RISSER 1982). It is a replication-defective acute tranforming virus. The P160 strain of A-MuLV has been molecularly cloned and analyzed in detail (LATT et al. 1983). At the 5′ end A-MuLV contains 1.3 kb of Mol-MuLV sequences and at the 3′ end 0.7 kb of sequences homologous to the 3′ end of Mol-MuLV. The v-*abl* sequences in the middle of the genome consist of a contiguous 4.3-kb segment. The gene in this virus therefore is 5′-*Δgag-abl-Δ* env 3′. Toward the C-terminus the v-*src*, v-*fps*, v-*yes*, and v-*fes* protein sequences exhibit significant homology (SCHWARTZ et al. 1983; SHIBUYA and HANAFUSA 1982; KITAMURA et al. 1982; HAMPE et al. 1982). Homology exists also between these oncogene products and cyclic AMP-dependent protein kinase (BARKER and DAYHOFF 1982). Recently the nucleic acid sequence of v-*abl* has been found to display homology with the above-mentioned v-*oncs* (BALTIMORE et al., unpublished data). It appears, then, that *abl* is yet another member of the protein kinase gene family. Interestingly, however, in A-MuLV the homologous sequences are located toward the N-terminus of the v-*abl* sequences and not at the C-terminus as with v-*src*, v-*fps*, v-*yes*, and v-*mos*. It is possible that the additional C-terminal sequences of *abl* which are not found in other v-*onc* products contribute to the unique biological properties of the *gag-abl* gene product.

A-MuLV virus nonproducer cells synthesize a genome size virus-specific RNA and no subgenomic RNAs. An A-MuLV specific protein can be detected by immunoprecipitation of labeled proteins from A-MuLV infected cells with antisera specific for the MuLV *gag* proteins p15, p12, and p30 (WITTE et al. 1978; REYNOLDS et al. 1978). With the P160 strain of A-MuLV this protein is 160000 daltons (ROSENBERG et al. 1980; GOFF et al. 1981; GRUNWALD et al. 1982). The genome organization of A-MuLV predicts that the gag-polyprotein is a *gag-abl* fusion protein. The *gag-abl* protein of the transformation-competent A-MuLV strains exhibit an in vitro protein kinase activity which is specific for tyrosine residues similar to that found to be associated with the *gag-fes* proteins of the ST- and the GA-FeSV. Both the *gag-abl* protein and immunoglobulin heavy chain are phosphorylated in the in vitro phosphorylation reaction (WITTE et al. 1980; PONTICELLI et al. 1982). A transformation-defective mutant of A-MuLV, P92td A-MuLV, lacks the protein kinase activity, suggesting that this unique activity is closely associated with A-MuLV mediated fibroblast transformation (ROSENBERG et al. 1980; REYNOLDS et al. 1980). Different genetic variants exist of A-MuLV; they are all thought to derive from the P160 strain of A-MuLV (ROSENBERG and WITTE 1980). The genomes of these variants have been shown to contain large deletions in the v-*abl* sequences or small genetic alterations which result in early termination of translation (GOFF et al. 1981). It is interesting that 60–70000 daltons of *abl* sequences of the P160 *gag-abl* protein can be deleted without significantly affecting the biological properties of the virus.

A-MuLV was isolated from a lymphosarcoma of a Balb/c mouse which had been treated with the steroid hormone prednisolone and infected with Moloney murine leukemia virus. (Mol-MuLV). In vivo A-MuLV induces hematopoietic neoplasms of the B-, T-, and possibly myeloid cell lineages (SKLAR et al. 1975; RASCHKE et al. 1978; COOK 1982; RISSER 1982). In this regard it is important to point out that no fibrosarcomas have been observed in animals infected with A-MuLV (ROSENBERG, unpublished data). Cells from target tissues have been transformed in vitro (ROSENBERG et al. 1975; ROSENBERG and BALTIMORE 1976). Susceptible cells were found in fetal liver and bone marrow and spleen of adult mice. In vitro transformed cells are early stage cells of the B-lymphocyte lineage. Furthermore, A-MuLV was shown to stimulate the proliferation of erythroid precursor cells and their differentiation in 8- to 10-day-old fetal livers (WANECK and ROSENBERG 1981). A-MuLV transforms NIH/3T3, Balb/3T3, NRK rat cells, and CCL64 mink cells in vitro. Other established cells lines and mouse embryo fibroblasts, however, are not transformed by this virus. Unstable transformation by A-MuLV is seen in certain sublines derived from Balb/3T3 cells (ZIEGLER et al. 1981). It is thought that the *gag-abl* transformation protein exhibits a lethal effect in these cells. Mutants of A-MuLV which stably transform the A-MuLV sensitive Balb/3T3 cells were found to contain deletions in the C-terminal domain of the *gag-abl* protein. Interestingly, the protein kinase activity of the mutant *gag-abl* protein and the efficiency of the mutant viruses in transforming lymphoid cells in vitro were comparable to those of wt A-MuLV.

In contrast to A-MuLV, the HZ2-FeSV was isolated from a multicentric fibrosarcoma and initial in vivo experiments indicate that the same type of neoplasm is induced upon infection of kittens with the virus. It appears, then, that A-MuLV and HZ2-FeSV transform different in vivo targets. As far as has been investigated, the in vitro transformation properties of the viruses resemble each other. However, it is not known whether the HZ2-FeSV exhibits a lethal effect and whether this virus transforms hematopoietic targets in vitro. Structurally the HZ2-FeSV and the A-MuLV genome resemble each other. Similarly to A-MuLV, the HZ2-FeSV specifies a *gag-abl* fusion protein with an associated tyrosine-specific protein kinase activity. However, it is not known precisely how much of the FeLV *gag* gene is retained in this virus, what the size of the v-*abl* insert is, what segment of the c-*abl* gene the HZ2-FeSV v-*abl* sequences are corresponding to, and whether the cat v-*abl* contains deletions or mutations. The fact that an *abl*-containing acute transforming retrovirus induces neoplasms of the fibroblast cell lineage is intriguing. The determination of the basis of the differing biology of A-MuLV and HZ2-FeSV will very likely further our understanding of *abl*-mediated cell transformation and possibly shed light on the role of the c-*abl* gene product in cell proliferation and differentiation.

3.5 v-*sis* ATVs: PI-FeSV

The PI-FeSV, even though isolated quite sometime ago, has only recently been characterized (Irgens et al. 1973; Besmer et al. 1983a). It is defective for replication and needs a helper virus for its reproduction. The integrated PI-FeSV provirus in cellular DNA of transformed virus nonproducer cells was characterized by Southern blot analysis. The v-*onc* sequences of PI-FeSV were found to be homologous with the v-*sis* sequences of the Woolly monkey derived Simian sarcoma virus (Besmer et al. 1983a). At the 5′ end the genome of this virus contains sequences homologous to FeLV 5′ sequences, including FeLV LTR and coding sequences for the FeLV *gag* proteins p15, p12, and p30. At the 3′ end it contains FeLV *env* sequences. The v-*sis* sequences are joined to FeLV *gag* sequences at their 5′ end and at the 3′ end v-*sis* is joined to sequences 5′ of the *pol-env* junction of FeLV, similarly to the ST-, the GA-, and the SM-FeSV. The gene order in PI-FeSV, then, appears to be 5′-*Δgag-sis-env*3′. The size of the PI-FeSV v-*sis* insert is not, however, known at present (Fig. 2).

Cells nonproductively infected by PI-FeSV express a 6.8-kb genomic RNA and a 3.3-kb subgenomic RNA. The 6.8-kb RNA contains 5′ *gag*, *sis*, and *env* sequences and lacks sequences of the FeLV *pol* gene. The 3.3-kb RNA only contains FeLV *env* sequences. Virus nonproducer cells synthesize a 76000-dalton with determinants of the FeLV *gag* proteins p15 and p30. This 76000-dalton protein does not exhibit any protein kinase activity in immunoprecipitates, nor is there an elevated level of phosphotyrosine in cells infected by the virus (Snyder, unpublished data). The arrangement

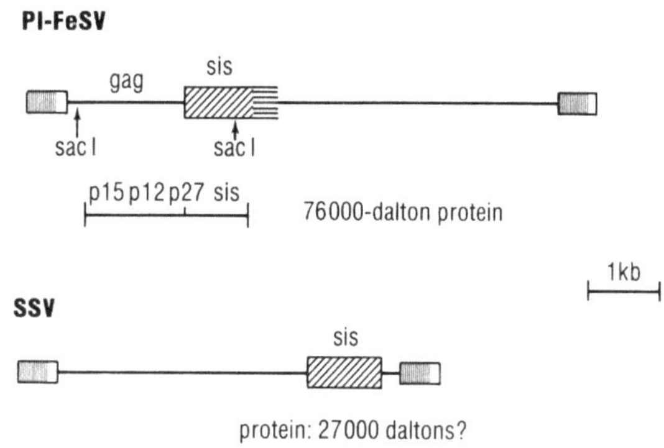

Fig. 2. Physical maps of PI-FeSV and SSV. The positions of v-*sis* sequences are indicated by cross-hatched boxes

of the *sis* sequences in the PI-FeSV genome predicts that the 76000-dalton *gag*-polyprotein is a *gag-sis* fusion protein.

Simian sarcoma virus (SSV) was generated by recombination of Simian sarcoma associated virus and woolly monkey c-*sis* sequences (THEILEN et al. 1971; ROBBINS et al. 1982; WONG-STAAL et al. 1981). The SSV DNA provirus has been molecularly cloned and extensively characterized (ROBBINS et al. 1981; GELMAN et al. 1981). The RNA genome of this virus is 6 kb long and the v-*sis* sequences are located toward the 3′ end of the viral genome. Sequence analysis of the entire SSV genome revealed the structure 5′ *gag-Δpol-Δ-env-sis-Δenv*3′. Thus, the *sis* sequences are joined to 5′ *env* sequences and this then predicts that the SSV transforming protein contains *env*-derived amino acids at the N-terminus and therefore is an *env-sis* fusion protein. The open reading frame furthermore predicts a 28000 dalton protein (ROBBINS et al. 1982). At the 3′ end the SSV v-*sis* sequences contain a 350 nucleotide untranslated region (DEVARE et al. 1983). No sequence similarities have been detected between v-*sis* and the other known retroviral oncogenes. V-*sis* therefore is not a member of the tyrosine kinase or the *ras* gene families. A comparison of the nucleic acid sequences of SSV and of the SSAV *env* gene with Moloney-MuLV revealed a striking similarity. This is not too surprising, since it is believed that the Gibbon leukemia virus with which SSAV is closely related has emerged from a murine retrovirus by cross-species infection (LIEBER et al. 1975). Similar to the Gibbon leukemia virus, FeLV also is proposed to derive from an ancestor of the murine leukemia virus (BENVENISTE et al. 1975). The transduction of cat c-*sis* and woolly monkey c-*sis* therefore involved evolutionarily related retroviruses.

In feline PI-FeSV and in SSV the *sis* protein sequences are expressed in different contexts of the retroviral genome: in PI-FeSV as a *gag-sis* protein and in SSV as an *env-sis* protein. In most acute transforming retroviruses which contain homologous *onc* sequences these sequences are expressed in analogous fashion (GRAF and STEHELIN 1982). PI-FeSV and SSV thus provide an example where homologous oncogenes are expressed with different strategies. Similarly, different strategies have been reported for the expression of the v-*myb* gene in the avian retroviruses AMV and E26 (BISTER et al. 1982).

The tissue culture characterization of PI-FeSV showed that PI-FeSV and SSV display indistinguishable behavior (BESMER et al. 1983a). The same characteristic focus morphology is observed with transformation induced by SSV and PI-FeSV, and mink CCL64 cells supported the replication of both viruses but were not transformed by either (MURPHY and BESMER, unpublished). SSV has been shown to induce well-differentiated fibrosarcomas and fibromas in marmosets when administered intramuscularly and gliomas when injected intracerebrally (WOLFE et al. 1971; WOLFE and DEINHARDT 1978). Only fibrosarcomas were obtained upon injection of kittens with PI-FeSV; however, no intracerebral inoculations have been performed with this virus (IRGENS et al. 1973). Thus whatever the mode of expression, the v-*sis* product appears to act similarly in both *SSV* and PI-FeSV.

3.6 GR-FeSV

The GR-FeSV, another replication-defective feline ATV, was isolated from an FeLV-associated fibrosarcoma of a cat (RASHEED et al. 1982). In vitro GR-FeSV transforms a large variety of mammalian cell line. Transformed GR-FeSV mink nonproducer cells have been used for the characterization of the translational products of the genome of this virus. Virus nonproducer cells synthesize a 70000-dalton protein with determinants of the FeLV *gag* protein p15. No proteins containing determinants of the structural FeLV proteins p30 and gp70 were detected in these cells. When assayed for in vitro protein kinase activity in immunoprecipitates, the GR-P70 protein exhibits kinase activity (NAHARRO et al. 1983). The P70 protein as well as immunoglobulin heavy chain was phosphorylated, and phosphorylation was shown to be specific for tyrosine. The level of phosphotyrosine-containing proteins in transformed cells has not been investigated yet. Since none of the FeLV gene products exhibit a tyrosine-specific protein kinase activity, it is likely that the GR-P70 is a *gag-onc* fusion protein. In an effort to determine whether the v-*onc* sequences of the GR-FeSV are related to v-*oncs* of other acute transforming retroviruses, the integrated GR-FeSV provirus was characterized by Southern blot analysis (NAHARRO et al. 1983). It thus was found that the GR-FeSV provirus does not contain nucleic acid sequences related to v-*fes*, v-*fms*, v-*abl*, v-*sis*, v-*bas*, v-*src*, and v-*myc*. Sequence homology with v-*yes*, v-*ros*, v-*erb*, v-*myb*, and v-*fos*, however, has not yet been investigated. The preliminary characterization of GR-FeSV indicates that this virus very likely contains yet another feline retroviral oncogene, which belongs to the tyrosine kinase gene family.

4 Origins of Feline ATVs

c-oncs. v-*onc* specific probes of ATVs have been used for the identification and characterization of sequences in the DNA of normal uninfected cells which are homologous to v-*oncs*. Cellular homologues of v-*onc* genes have been designated c-*onc* (Coffin et al. 1981). C-*onc* genes are highly conserved in vertebrate species and some of them have even been detected in *Drosophila* using low-stringency hybridization conditions (Shilo and Weinberg 1981). Most v-*onc* genes are related to c-*onc* genes in the same way as large cellular genes are related to their mRNAs, that is, c-*onc* genes contain intervening sequences. v-*oncs* therefore appear to be DNA copies of segments of c-*onc* mRNAs. The expression of many c-*oncs* in normal cells was found to be tissue specific; others, however, are expressed similarly in cells of different tissues (Shibuya et al. 1982; Gonda et al. 1982; Müller et al. 1982). No relation is apparent between the expression of c-*oncs* in different tissues and the in vivo transformation specificities of corresponding v-*onc* containing ATVs. Protein products of some c-*oncs* have been identified, but their function in normal cells has so far remained elusive. If retroviral oncogenes contain activities found in normal cells, then how and why do these viral gene products transform infected cells? Two hypothesis have been put forward: (a) an overproduction of a normal cell product which is under the control of the strong viral promoter of the ATV may lead to transformation; (b) v-*oncs* may be altered c-*oncs*. As a consequence, altered v-*onc* protein products could have acquired different substrate specificities which could be responsible for the transformation of the target cells. The gene dosage theory has received support from experiments by Van de Woude and collaborators, in which a viral LTR fused to c-*mos* induced transformation in fibroblats (Blair et al. 1981), and from experiments by Hayward and coworkers, in which the ALV LTR was found to amplify the expression of c-*myc* in bursal lymphomas of chickens by downstream promotion (Hayward et al. 1981) or by enhancement (Payne et al. 1982). More recently evidence has also been obtained which might support the mutational theory of transformation (Takeya and Hanafusa 1983). When the Rous sarcoma virus v-*src* nucleic acid sequence was compared with the sequence of the chicken c-*src*, it was found that the last 19 carboxy terminal amino acids of pp60 c-*src* in v-*src* are replaced by a new set of 12 amino acids which derive from sequences beyond the termination codon of the c-*src* coding region. This indicates that the RSV pp*src* protein is an altered c-*src* protein. The analysis of c-*onc* genes and their products, and the determination of their relationship to the viral oncogenes, are clearly of importance for the elucidation of how retroviral oncogenes transform cells. The structure and the expression of c-*oncs* has been studied mostly in chickens, mice, and humans. Except for c-*fes*, not very much is known about c-*oncs* and their products in cats.

4.1 Feline c-*oncs*: c-*fes*

The cat c-*fes* locus was investigated by restriction mapping of normal cat cell DNA (Franchini et al. 1981). Different v-*fes* hybridization probes were used for the identification of restriction fragments which contain *fes* sequences. Only one restriction fragment was found to hybridize to the ST-FeSV v-*fes* hybridization probe when the DNA was cleaved with the restriction enzymes Eco R1, HindIII, and Bgl II, indicating that the gene was unique and present once per haploid genome. The sequences homologous to the ST-FeSV v-*fes* probe were found to be interspersed by at least three segments of sequences which are not homologous to v-*fes*, indicating that c-*fes* contains intervening sequences. The v-*fes* homology was thus spread over 4.5 kb of DNA. Since v-*fes* very likely represents only a segment of the c-*fes* mRNA, it is to be expected that the cat c-*fes* gene is significantly larger than 4.5 kb. Recently the human c-*fes* locus has been molecularly cloned by two groups (Franchini et al. 1982; Groffen et al. 1982). The analysis of the human c-*fes* DNA clones revealed striking structural similarities with the cat c-*fes* gene. The investigation of RNA expression of c-*fes* has been difficult because of the low levels of expression of c-*fes* mRNA (Frankel et al. 1979). Thus the size of the c-*fes* mRNA is not known. v-*fes* specific hyperimmune sera had been made against both the v-*fes* determinants of the GA-FeSV *gag-fes* and the ST-FeSV *gag-fes* proteins. These *fes* antisera were used in order to identify c-*fes* protein products in normal cells (*fes* NCPs). A 92000-dalton phosphoprotein was found in feline embryofibroblasts as well as in the lymphoid FL74 cell line (Barbacid et al. 1981). A cross-reactive protein was also detected in other mammalian cells. No detailed characterization of the feline *fes*-NCP is available yet.

The c-*fps* gene products in chickens have been investigated more thoroughly than the homologous c-*fes* gene products in cats. c-*fps* RNA was found to be expressed at very low levels in many tissues, but not in bone marrow (2.5 copies per cell) and lung (1.1 copy per cell), indicating that c-*fps* expression is tissue specific (Shibuya et al. 1982). Using *fps*-specific hyperimmune sera, a 98000-dalton normal cell protein (NCP) was detected in bone marrow cells (Mathey-Prevot et al. 1982). Tryptic peptide analysis of the 98000-dalton NCP showed that this protein was structurally related to the FSV *gag-fps* protein. In vivo the NCP 95 is a phosphoprotein and in vitro it exhibits a protein kinase activity similar to that of the *gag-fps* transforming protein of FSV. Phosphoamino acid analysis of in vivo labeled NCP revealed phosphoserine. The in vitro labeled protein, on the other hand, contained phosphotyrosine. In contrast, the in vivo labeled *gag-fps* protein contained both phosphotyrosine and phosphoserine. The significance of this difference is not understood.

4.2 Mechanism of Transduction

The mechanism by which retroviruses transduce cellular oncogenes is not very well understood. Several mechanisms have been proposed for their origin. One model proposes that recombination occurs at the RNA level. A second model proposes that the recombination event occurs entirely at the DNA level, and in a third model recombination at both the DNA and RNA level is proposed. In the RNA recombination mechanism c-*onc* mRNA would be packaged into virus particles to generate heterozygous virus progeny and recombination is proposed to occur in the following replication cycle during reverse transcription by template switching (COFFIN 1979). A mutant of Rous sarcoma virus is known which packages cellular mRNAs very efficiently. In general though, cellular mRNAs are not packaged significantly into virus particles. In the DNA recombination model a DNA provirus would integrate upstream of a c-*onc* gene. A deletion (or DNA rearrangement) then would fuse an internal viral segment with the c-*onc* gene, deleting the 3′ half of the provirus and the 5′ end of the c-*onc* gene. A second provirus is proposed to integrate downstream of the c-*onc* gene and an analogous deletion would fuse 3′ c-*onc* sequences with 3′ viral sequences. Transcription and splicing of c-*onc* exons would then generate the RNA genome of the recombinant virus. In the DNA-RNA recombination model, a DNA provirus integrates upstream of the c-*onc* gene and deletion then, as in the DNA model, would fuse internal viral sequences to c-*onc* sequences. Upon transcription from the viral promoter and splicing, a hybrid virus-c-*onc* mRNA would be obtained which contains the viral packaging site. The hybrid RNA could thus be packaged efficiently into virus particles along with helper virus RNA and the second recombination could occur in the next replication cycle during reverse transcription by template switching. In order to investigate the recombination events in the generation of acute transforming retroviruses, sequences at the virus-v-*onc* junctions have been compared with the corresponding c-*onc* sequences and the sequences of the parental helper virus. Recent analysis of the nucleotide sequence of avian myeloblastosis virus and of c-*myb* revealed that the v-*myb* sequences at the 5′ recombination site derive from c-*myb* intron sequences (KLEMPNAUER et al. 1982). With Rous sarcoma virus the v-*src* sequences at the 5′ recombination site were found to be missing in the c-*src* mRNA (TAKEYA and HANAFUSA 1983). Both results suggest that these 5′ recombination sites were generated by recombination events at the DNA level. Nucleotide homologies of 2, 3, 4, and 6 nucleotides with AMV, Moloney MSV, Abelson MuLV, and Simian Sarcoma Virus have been observed between helper virus and c-*onc* sequences at the recombination sites (KLEMPNAUER et al. 1982; VAN BEVEREN et al. 1981; JOSEPHS et al. 1983; WANG et al., unpublished data). Similar homologies are often found at deletion end points in prokaryotic DNA (ALBERTINI et al. 1982). Recent nucleotide sequence analysis of the St- and the GA-FeSV revealed that the 3′ recombination sites in these two viruses are identical (HAMPE et al. 1982). If these viruses have arisen independently, as is assumed, then their 3′ recombination points must contain common features. It seems possible, for example, that an extensive homology exists between c-*fes* and the parental FeLV helper virus sequences at the recombination site, similar to the homologies found in deletion hot spots of bacteria, if recombination occurs at the DNA level.

A precedent for the RNA recombination step in the DNA-RNA recombination model has been described (GOLDFARB and WEINBERG 1981). Upon transfection of NIH 3T3 cells with unintegrated Harvey sarcoma virus DNA, transformants were obtained which contained the 5′ half of the Harvey sarcoma virus genome fused to cellular sequences. Upon infection of these transformants with Moloney-MuLV, transforming virus was rescued at a low frequency in which the missing 3′ viral sequences had been reacquired by illegitimate recombination from the Moloney-MuLV helper virus.

In both the DNA-DNA and the DNA-RNA recombination models the transduced c-*onc* gene gets transcriptionally activated. It is known that the Abelson-MuLV transforming protein exhibits a lethal effect in certain cell types. Transduction of c-*onc* genes thus may only occur in cells where transcriptional activation does not affect their viability.

From the evidence above it appears that at least one recombination step occurs at the DNA level. The DNA-RNA recombination model is favored by most people because the DNA-DNA model requires two specific proviral integrations and deletions and therefore is less likely to occur than the DNA-RNA scheme, which only requires one (Klempnauer et al. 1982).

4.3 How Many Retroviral Oncogenes Are There?

To date at lcast 16 different retroviral oncogenes are known to exist in about 35 viruses isolated from different species (Weiss et al. 1982). In feline ATVs five different v-*oncs* have been found in only seven virus strains. The high ratio of the number of v-*oncs* to the number of virus isolates suggests that there is no significant specificity for the transduction cellular oncogenes by FeLVs, with the exception of c-*fes*. The characterization of more feline ATVs very likely will reveal new transducible feline c-*oncs*. Interestingly, three of the four characterized feline retroviral oncogenes are homologous with v-*oncs* of viruses isolated from other species, demonstrating that retroviruses transduce homologous cellular genes from different species. The increasing number of oncogenes which are shared by viruses isolated from the same species as well as viruses isolated from different species predicts that the number of cellular oncogenes which can be transduced by retroviruses is limited and possibly will not exceed 20–30. More ATVs will undoubtedly have to be characterized to substantiate these predictions.

Acknowledgments. I would like to thank Peter Bergold, William Hardy, Naomi Rosenberg, Charles Sherr, Harry Snyder, and the Viral Oncology group for many helpful discussions, Nancy Famulari and Allen Silverstone for thoughtful criticism of this manuscript, and Atlanta Chegee for typing it. Unpublished work cited was supported by grants MV 82 from the American Cancer Society and CA-32926 and CA-08748 from the National Cancer Institute.

References

Abelson HT, Rabstein LS (1970a) Influence of prednisolone on Moloney leukemogenic virus in Balb/c mice. Cancer Res 30:2208–2212
Abelson HT, Rabstein LS (1970b) Lymphosarcoma: virus induced thymus independent disease in mice. Cancer Res 30:2213–2222
Albertini AMM, Hofer MP, Calos, Miller JM (1982) On the formation of spontaneous deletions: The importance of short sequence homologies in the generation of large deletions. Cell 29:319–328
Anderson SJ, Furth M, Wolff L, Ruscetti SK, Sherr CJ (1982) Monoclonal antibodies to the transformation-specific glycoprotein encoded by the feline retroviral oncogene v-*fms*. J Virol 44:696–702
Baluda MA, Roy-Burman P (1973) Partial characterization of RD-114 virus by DNA-RNA hybridization studies. Nature 244:59–62
Barbacid M (1981) Cellular transformation by subgenomic feline sarcoma virus DNA. J Virol 37:518–523
Barbacid M, Lauver AL, Devare SG (1980a) Biochemical and immunological characterization of polyproteins coded for by the McDonough, Gardner-Arnstein and Snyder-Theilen strain of feline sarcoma virus. J Virol 33:196–207
Barbacid M, Beemon K, Devare SG (1980b) Origin and functional properties of the sarc gene product of the Snyder-Theilen strains of feline sarcoma virus. Proc Natl Acad Sci USA 77:5158–5162
Barbacid M, Donner L, Ruscetti SK, Sherr CJ (1981) Transformation defective mutants of Snyder-Theilen feline sarcoma virus lack tyrosine specific kinase activity. J Virol 39:246–254
Barker WC, Dayhoff MO (1982) Viral src gene products are related to the catalytic chain of mammalian cAMP-dependent protein kinase. Proc Natl Acad Sci USA 79:2836–2839
Benveniste RE, Sherr C, Todaro G (1975) Evolution of type C viral genes: origin of feline leukemia virus. Science 190:886–888
Besmer P, Snyder HW, Murphy JR, Hardy WD, Parodi A (1983a) The Parodi-Irgens feline sarcoma virus and simian sarcoma virus have homologous oncogenes. J Virol 46:606–613
Besmer P, Hardy WD Jr, Zuckerman E, Lederman L, Snyder HW (1983b) The Hardy-Zuckerman 2FeSV and Abelson MuLV have homologous oncogenes. Nature
Bister K, Nunn M, Moscovici C, Perbal B, Baluda MA, Duesberg PH (1982) Acute leukemia virus E26 and avian myeloblastosis virus have related transformation specific RNA sequences but different genetic structures, gene products and oncogenic properties. Proc Natl Acad Sci USA 79:3677–3681
Blair DG, Oskarsson M, Wood TG, McClements WL, Fischinger PJ, Van de Woude GF (1981) Activation of the transforming potential of a normal-cell sequence, a molecular model for oncogenesis. Science 212:941–943
Blomberg J, Van de Ven WJM, Reynolds FH, Nalewaik RP, Stephenson JR (1981) Snyder-Theilen feline sarcoma virus p85 contains a single phosphotyrosine acceptor site recognized by its associated protein kinase. J Virol 38:886–894
Boss M, Greaves M, Teich N (1979) Abelson virus-transformed hematopoietic cell lines with pre-B-cell characteristics. Nature 278:551–553
Chan EW, Schiob-Stansly PE, O'Connor TE (1974) Rescue of cell transforming virus from a non virus producing bovine cell culture transformed by feline sarcoma virus. J Natl Cancer Inst 52:469–472
Coffin JM (1979) Structure replication and recombination of retrovirus genomes: some unifying hypotheses. J Gen Virol 42:1–26
Coffin JM, Varmus HE, Bishop JM, Essex M, Hardy WD, Martin GS, Rosenberg NE, Scolnick EM, Weinberg RA, Vogt PK (1981) A proposal for naming host-cell derived inserts in retrovirus genomes. J Virol 40:953–957
Cook W (1982) Rapid thymomas induced by Abelson murine leukemia virus. Proc Natl Acad Sci USA 79:2917–2921
Cooper JA, Hunter T (1981) Four different classes of retroviruses induce phosphorylation of tyrosine present in similar cellular proteins. Mol Cell Biol 1:394–407

Deinhardt F, Wolfe LG, Theilen GH, Snyder SP (1970) ST-feline fibrosarcoma virus: induction of tumors in marmoset monkeys. Science 167:881
Devare SG, Reddy EP, Law JD, Robbins KC, Aaronson SA (1983) Nucleotide sequence of the simian sarcoma virus genome: Demonstration that its acquired cellular sequences encode the transforming gene product $p28^{sis}$. Proc Natl Acad Sci USA 80:731–735
Donner L, Turek LP, Ruscetti SK, Fedele LA, Sherr CJ (1980) Transformation defective mutants of feline sarcoma virus which express a product of the viral src gene. J Virol 35:129–140
Donner L, Fedele LA, Garon CF, Anderson SJ, Sherr CJ (1982) McDonough feline sarcoma virus: characterization of the molecularly cloned provirus and its feline oncogene (v-*fms*). J Virol 41:489–500
East JL, Knesek JE, Allen PT, Domchovsky L (1973) Structural characteristics and sequence analysis of genomic RNA from RD-114 virus and feline RNA tumor viruses. J Virol 12:1085–1091
Even J, Anderson SJ, Hampe A, Galibert F, Long D, Khoury GK, Sherr CJ (1983) Mutant feline sarcoma proviruses containing the viral oncogene (v-*fes*) and either feline or murine control elements. J Virol 45:1004–1016
Fedele LA, Even J, Garon CF, Donner L, Sherr CJ (1981) Recombinant bacteriophages containing the integrated transforming provirus of Gardner-Arnstein feline sarcoma virus. Proc Natl Acad Sci USA 78:4036–4040
Franchini G, Even J, Sherr CJ, Wong-Staal F (1981) *Onc*-sequences (v-*fes*) of Snyder-Theilen feline sarcoma virus are derived from noncontiguous regions of a cat cellular gene (c-*fes*). Nature 290:154–157
Franchini G, Gelman EP, Dalla-Favera R, Gallo RC, Wong-Staal F (1982) Human gene (c-*fes*) related to the *onc* sequences of Snyder-Theilen feline sarcoma virus. Mol Cell Biol 2:1014–1019
Frankel AE, Gilbert JH, Porzig KT, Scolnick EM, Aaronson SA (1979) Nature and distribution of feline sarcoma virus nucleotide sequences. J Virol 30:821–827
Gardner MB, Rongey RW, Arnstein P, Estes JD, Sarma P, Huebner RJ, Rickard CJ (1970) Experimental transmission of a feline fibrosarcoma to cats and dogs. Nature 226:807–809
Gelman EP, Wong-Staal F, Kramer RA, Gallo RC (1981) Molecular cloning and comparative analysis of the genomes of Simian sarcoma virus and its associated helper virus. Proc Natl Acad Sci USA 78:3373–3378
Goff SP, Gilboa E, Witte ON, Baltimore D (1980) Structure of the Abelson murine leukemia virus genome and the homologous cellular gene: Studies with cloned viral DNA. Cell 22:777–785
Goff SP, Witte ON, Gilboa E, Rosenberg W, Baltimore D (1981) Genome structure of Abelson murine leukemia virus variants: proviruses in fibroblasts and lymphoid cells. J Virol 38:460–468
Goldfarb, MP, Weinberg RA (1981a) Structure of the provirus within NIH3T3 cells transfected with Harvey sarcoma virus DNA. J Virol 38:125–135
Goldfarb MP, Weinberg RA (1981b) Generation of novel, biologically active Harvey sarcoma virus via apparent illegitimate recombination. J Virol 38:136–150
Gonda TJ, Sheiness DK, Bishop JM (1982) Transcripts from the cellular homologs of retroviral oncogenes: Distribution among chicken tissues. Mol Cell Biol 2:617–624
Graf T, Stehelin D (1982) Avian leukemia viruses and genome structure. Biochem Biophys Acta 651:245–271
Groffen J, Heisterkamp N, Grosveld F, Van de Ven W, Stephenson JR (1982) Isolation of human oncogene sequences (v-*fes* homolog) from a cosmid library. Science 261:1136–1138
Grunwald DJ, Dale B, Dudley J, Lamph W, Sugden B, Ozanne B, Risser R (1982) Loss of viral gene expression and retention of tumorigenicity by Abelson lymphoma cells. J Virol 43:92–103
Hampe A, Laprevotte I, Galibert F, Fedele LA, Sherr CJ (1982) Nucleotide sequences of feline retroviral oncogenes (v-fes) provide evidence for a family of tyrosine specific protein kinase genes. Cell 30:775–785

Hampe A, Gobet M, Even J, Sherr CJ, Galibert F (1983) Nucleotide sequences of feline sarcoma virus long terminal repeats and 5′ leaders show extensive homology to those of other mammalian retroviruses. J Virol 45:466–472
Hardy WD Jr (1981) The biology and virology of FeSV. In: Hardy WD Jr, Essex M, McClelland AJ (eds) Feline leukemia virus. Elsevier/North Holland, New York, pp 79–118
Hardy WD Jr, Old LJ, Hess PW, Essex M, Cotter S (1973) Horizontal transmission of feline leukemia virus. Nature 244:266
Hardy WD, Zuckerman E, Markovich R, Besmer P, Snyder HW Jr (1982) In: Yohn DS, Blakeslee JR (eds) Advances in comparative leukemia research 1981. Elsevier/North Holland, New York, pp 205–206
Hayward WS, Neel BG, Astrin SM (1981) Activation of a cellular onc gene by promoter insertion in ALV induced lymphoid leukosis. Nature 290:475–480
Henderson IC, Lieber MM, Todaro GJ (1974) Mink cell line Mv/Lv (CCL64) Focus formation and the generation of nonproducer transformed cell lines with murine and feline sarcoma viruses. Virology 60:282–287
Irgens K, Wyers M, Moraillon A, Parodi A, Fortuny V (1973) Isolement d'un fibrosarcome spontané du chat: étude du pouvoir sarcomatogène *in vivo*. CR Acad Sci Paris 276:1783–1786
Jarrett O (1980a) Natural occurrence of subgroups of feline leukemia virus. Cold Spring Harbor Conf. Cell Proliferation 8:603–612
Jarrett O (1980b) Feline leukemia virus subgroups. In: Hardy WD et al. (eds) Feline leukemia virus. Elsevier/North Holland, New York, pp 473–479
Jarrett O, Russell PH (1978) Differential growth and transmission in cats of FeLVs of subgroups A and B. Int J Cancer 21:466–472
Jarrett HFH, Jarrett O, Mackey L, Laird H, Hardy WD Jr, Essex M (1973a) Horizontal transmission of leukemia virus and leukemia in the cat. J Natl Cancer Inst 51:833–841
Jarrett O, Laird HM, Hay D (1973b) Determinants of the host range of feline leukemia viruses. J Gen Virol 20:169–175
Jarrett O, Hardy WD Jr, Golder MC, David H (1978) The frequency of occurrence of feline leukemia virus subgroups in cats. Int J Cancer 21:334–337
Josephs SF, Dalla-Favera R, Gelman EP, Gallo RC, Wong-Staal F (1983) 5′ viral and human cellular sequences corresponding to the transforming gene of Simian sarcoma virus. Science 219:503–505
Kahn AS, Deobagkar DN, Stephenson JR (1978) Isolation and characterization of a feline sarcoma virus-coded precursor polyprotein. J Biol Chem 253:8894–8901
Kitamura N, Kitamura A, Toyoshima K, Hirayama Y, Yoshida M (1982) Avian sarcoma virus Y73 genome sequence and structural similarity of its transforming gene product to that of Rous sarcoma virus. Nature 297:205–208
Klempnauer KH, Gonda TJ, Bishop JM (1982) Nucleotide sequence of the retroviral leukemia gene v-*myb* and its cellular progenitor c-*myb*: the architecture of a transformed oncogene. Cell 31:453–463
Latt SA, Goff SP, Tabin CJ, Paskind M, Wang JYJ, Baltimore D (1983) Cloning and analysis of reverse transcript P160 genomes of Abelson murine leukemia virus. J Virol 45:1195–1199
Levin R, Ruscetti SK, Parks WP, Scolnick EM (1976) Expression of feline type-c virus in normal and tumor tissues of the domestic cat. Int J Cancer 18:661
Lieber MM, Sherr CJ, Todaro GJ, Benveniste RE, Callahan R, Coon HG (1975) Isolation from the asian mouse *Mus caroli* of an endogenous type-C virus related to infectious primate type c viruses. Proc Natl Acad Sci USA 72:2315–2319
Livingston D, Todaro GJ (1973) Endogenous type C virus from a cat cell clone with properties distinct from previously described feline type C virus. Virology 53:142–151
Mathey-Prevot B, Hanafusa H, Kawai S (1982) A cellular protein is immunologically crossreactive with and functionally homologous to Fujinami sarcoma virus transforming protein. Cell 28:897–906
McAllister RM, Michelson M, Gardner MB, Rongey RW, Rasheed S, Sarma PS, Huebner RJ, Hatnaka M, Oroszlan S, Gilden RV, Kabigting A, Vernon L (1972) C-type virus released from cultured simian rhabdomyosarcoma cells. Nature 235:3–6

McCullough B, Schaller J, Shadduck JA, Yohn DS (1972) Induction of malignant melanomas associated with fibrosarcoma in gnotobiotic cats inoculated with Gardner feline fibrosarcoma virus. J Natl Cancer Inst 48:1893–1896
McDonough SK, Larsen S, Brodey RS, Stock WD, Hardy WD Jr (1971) A transmissible feline fibrosarcoma of viral origin. Cancer Res 31:953
Müller R, Salomon DJ, Tremblay JM, Cline MJ, Verma JM (1982) Differential expression of cellular oncogenes during pre- and postnatal development of the mouse. Nature 299:640–644
Mullins JI, Casey JW, Nicolson MD, Burck KB, Davidson N (1981) Sequence arrangement and biological activity of cloned feline leukemia proviruses from a virus productive human cell line. J Virol 38:688–703
Naharro G, Dunn CY, Robbins KC (1983) Analysis of the primary translational product and integrated DNA of a new feline sarcoma virus, GR-FeSV. Virology 125:502–507
Neil JC, Smart JE, Hayman MJ, Jarrett O (1980) Polypeptides of feline leukemia virus: A glycosylated *gag*-related protein is released into culture fluids. Virology 105:250–253
Niederkorn JY, Shadduck JA, Albert D, Essex M (1980) Antibody response to FOCMA in feline sarcoma virus induced ocular melanomas. In: Hardy WD Jr, Essex M, McClelland AJ (eds) Feline leukemia virus. Elsevier/North Holland, New York, pp 181–185
Okabe H, DuBuy J, Gilden RV, Gardner MB (1978) A portion of the feline leukemia genome is not endogenous in cat cells. Int J Cancer 22:70–78
Okazinski GF, Velicer LF (1976) Analysis of intracellular feline leukemia virus proteins. I. Identification of a 60,000 dalton precursor of FeLV p30. J Virol 20:96–106
Okazinski GF, Velicer LF (1977) Analysis of intracellular feline leukemia virus proteins: II. Generation of FeLV structural proteins from precursor polypeptides. J Virol 22:74–85
Patchinsky T, Hunter T, Esch FS, Cooper JA, Sefton BM (1982) Analysis of the sequence of amino acids surrounding sites of tyrosine phosphorylation. Proc Natl Acad Sci USA 79:973–977
Payne GS, Bishop JM, Varmus HE (1982) Multiple arrangements of viral DNA and an activated host oncogene in bursal lymphoma. Nature 295:209–214
Pinter A, Fleissner E (1979) Structural proteins of retroviruses: Characterization of oligomeric complexes of murine and feline leukemia virus envelope and core components formed upon crosslinking. J Virol 30:157–165
Ponticelli AS, Whitlock CA, Rosenberg N, Witte ON (1982) In vivo tyrosine phosphorylations of the Abelson virus transforming protein are absent in its normal cellular homolog. Cell 29:953–960
Porzig KJ, Barbacid M, Aaronson SA (1979) Biological properties and translational products of three independent isolates of feline sarcoma virus. Virology 92:91–107
Quintrell N, Varmus HE, Bishop JM, Nicholson MO, McAllister RM (1974) Homologies among the nucleotide sequences of the genomes of c-type viruses. Virology 58:568–575
Raschke WC, Baird S, Ralph P, Nakoina I (1978) Functional macrophage cell lines transformed by Abelson leukemia virus. Cell 15:261–267
Rasheed S, Gardner MB, Chan E (1976) Amphotropic host range of naturally occurring wild mouse leukemia viruses. J Virol 19:13–18
Rasheed S, Barbacid M, Aaronson SA, Gardner MB (1982) Origin and biological properties of a new feline sarcoma virus. Virology 117:238–244
Reynolds FH, Sacks TL, Deobaghar DN, Stephenson JR (1978) Cells nonproductively transformed by Abelson murine leukemia virus express a high molecular weight polyprotein containing structural and nonstructural components. Proc Natl Acad Sci USA 75:3974–3978
Reynolds FH, Van de Ven WJM, Stephenson JR (1980) Abelson murine leukemia virus transformation defective mutant with impaired P120 associated protein kinase activity. J Virol 36:374–386
Reynolds FH, Van de Ven WJM, Blomberg J, Stephenson JR (1981 a) Differences in mechanism of transformation by independent feline sarcoma virus isolates. J Virol 38:1084–1089
Reynolds FH, Van de Ven WJM, Blomberg J, Stephenson JR (1981 b) Involvement of a high molecular weight protein translational product of Snyder-Theilen feline sarcoma in malignant transformation. J Virol 37:643–653

Rho HM, Gallo RC (1979) Characterization of reverse transcriptase from feline leukemia virus by radioimmunoassay. Virology 99:192–196
Rickard CG, Post JE, Noronha F, Barr LM (1969) A transmissible virus induced lymphocytic leukemia of the cat. J Natl Cancer Inst 42:987
Risser R (1982) The pathogenesis of Abelson virus lymphomas of the mouse. Biochem Biophys Acta 651:213–244
Risser R, Potter M, Rowe WP (1978) Abelson virus induced lymphomagenesis in mice. J Exp Med 148:714–726
Robbins KC, Barbacid M, Porzig KJ, Aaronson SA (1979) Involvement of different exogenous feline leukemia virus subgroups in the generation of independent feline sarcoma virus isolates. Virol 97:1–11
Robbins KC, Devare SG, Aaronson SA (1981) Molecular cloning of integrated Simian sarcoma virus: genomic organization of infectious DNA clones. Proc Natl Acad Sci USA 78:2918–2922
Robbins KC, Hill RL, Aaronson SA (1982) Primate origin of the cell derived sequences of simian sarcoma virus. J Virol 41:721–725
Rosenberg N, Baltimore D (1976) A quantitative assay for transformation of bone marrow cells by Abelson murine leukemia virus. J Exp Med 143:1453–1463
Rosenberg N, Witte ON (1980) Abelson murine leukemia virus mutants with alterations in the virus specific P120 molecule. J Virol 33:340–348
Rosenberg N, Baltimore D, Scher CD (1975) In vitro transformation of lymphoid cells by Abelson murine leukemia virus. Proc Natl Acad Sci USA 72:1932–1936
Rosenberg NE (1982) Abelson leukemia virus. Curr Top Microbiol Immunol 101:95–126
Rosenberg NE, Clark DR, Witte ON (1980) Abelson murine leukemia virus mutants deficient in kinase activity and lymphoid cell transformation. J Virol 36:766–774
Rosenberg ZF, Crowther RL, Essex M, Jarrett O, Hazeltine WA (1981) Isolation via transfection of feline leukemia viruses from DNA of naturally occurring feline lymphoma. Virology 115:203–210
Ruscetti SK, Turek LP, Sherr CJ (1980) Three independent isolates of feline sarcoma virus code for three distinct *gag-x* polyproteins. J Virol 35:259–264
Sarma PS, Log T (1971) Viral interference in feline leukemia sarcoma complex. Virology 44:352–358
Sarma PS, Log T (1973) Subgroup classification of feline leukemia and sarcoma viruses by viral interference and neutralization tests. Virology 54:160–169
Sarma PS, Baskar JF, Gilden RV, Gardner MB, Huebner RJ (1971a) In vitro isolation and characterization of the GA strain of feline sarcoma virus. Proc Soc Biol Med 137: 1333–1336
Sarma SP, Log T, Theilen GH (1971b) ST feline sarcoma virus. Biological characteristics and in vitro propagation. Proc Soc Biol Med 137:1444–1448
Sarma PS, Sharary AL, McDonough S (1972) The SM strain of feline sarcoma virus. Biologic and antigenic characterization of virus. Proc Soc Biol Med 140:1365–1368
Sarma PS, Log T, Jain D, Hill PR, Huebner RJ (1975) Differential host range of viruses of feline leukemia sarcoma complex. Virology 64:438–446
Schwartz DE, Tizard R, Gilbert N (1983) Nucleotide sequence of Rous sarcoma virus. Cell 32:853–869
Scolnick EM, Park WP, Livingston D (1972) Radioimmunoassay of mammalian c-type proteins: species specific reactions of murine and feline viruses. J Immunol 109:570–577
Sherr CJ, Todaro GJ (1974) Radioimmunoassay of the major group specific protein of endogenous baboon type C viruses: Relation to the RD-114/CCC group and detection of antigen in normal baboon tissues. Virology 61:168–181
Sherr CJ, Fedele LA, Donner L, Turek LP (1979) Restriction endonuclease mapping of unintegrated proviral DNA of Snyder-Theilen feline sarcoma virus: localization of sarcoma specific sequences. J Virol 32:860–875
Sherr CJ, Fedele LA, Oskarsson M, Maizel J, Van de Woude G (1980a) Molecular cloning of Snyder-Theilen feline leukemia and sarcoma viruses: comparative studies of feline sarcoma virus with its natural helper virus and with Moloney murine sarcoma virus. J Virol 34:200–212

Sherr CJ, Donner L, Fedele LA, Turek L, Even J, Ruscetti SK (1980b) Molecular structure and products of feline sarcoma and leukemia viruses: relationship to FOCMA expression. In: Hardy WD et al. (eds) Feline leukemia virus. Elsevier/North Holland, New York, pp 293–307

Sherr CJ, Sen A, Todaro GJ, Sliski A, Essex M (1981) Pseudotypes of feline sarcoma virus contain an 85,000 dalton protein with feline oncornavirus-associated cell membrane antigen (FOCMA) activity. Proc Natl Acad Sci USA 75:1505

Shibuya M, Hanafusa H (1982) Nucleotide sequence of Fujinami sarcoma virus: evolutionary relationship of its transforming gene with transforming genes of other sarcoma viruses. Cell 30:787–795

Shibuya M, Hanafusa T, Hanfusa H, Stephenson JR (1980) Homology exists among the transforming sequences of the avian and feline sarcoma viruses. Proc Natl Acad Sci USA 77:6536–6540

Shibuya M, Hanafusa H, Balduzzi PC (1982) Cellular sequences related to three new onc genes of avian sarcoma virus (*fes*, *yes*, and *ros*) and their expression in normal and transformed cells. J Virol 42:143–152

Shilo BZ, Weinberg RA (1981) DNA sequences homologous to vertebrate oncogenes are conserved in *Drosophila melanogaster*. Proc Natl Acad Sci USA 78:6789–6792

Sklar MD, Shevach EM, Green I, Potter M (1975) Transplantation and preliminary characterization of lymphocyte surface markers of Abelson virus induced lymphomas. Nature 253:550–552

Snyder HW Jr (1982) Biochemical characterization of protein kinase activities associated with transforming gene products of the Snyder-Theilen and Gardner-Arnstein strains of feline sarcoma virus. Virology 117:165–172

Snyder HW Jr, Hardy WD Jr, Zuckerman EE, Fleissner (1978) Characterization of a tumor specific antigen on the surface of feline lymphosarcoma cells. Nature 275:656–658

Snyder SP (1971) Spontaneous feline fibrosarcoma: transmissibility and ultrastructure of associated virus like particles. J Natl Cancer Inst 47:1079–1085

Snyder SP, Theilen GH (1969) Transmissible feline fibrosarcoma. Nature 221:1074–1075

Stehelin D, Varmus HE, Bishop JM, Vogt PK (1976) DNA related to the transforming gene(s) of avian sarcoma viruses is present in normal avian DNA. Nature 260:140–143

Stephenson JR, Kahn AS, Sliski AH, Essex M (1977) Feline oncornavirus-associated cell membrane antigen: evidence for an immunologically cross-reactive feline sarcoma virus coded protein. Proc Natl Acad Sci USA 74:5608–5112

Takeya T, Hanafusa H (1983) Structure and sequence of the cellular gene homologous to the RSV src gene and the mechanism for generating the transforming virus. Cell 32:881–890

Theilen GH, Gould D, Fowler M, Dungworth DL (1971) C-type virus in tumor tissue of a woolly monkey (*Lagothrix* spp.) with fibrosarcoma. J Natl Cancer Inst 47:881–889

Todaro GJ, Benveniste RE, Callahan R, Lieber MM, Sherr CJ (1974) Endogenous primate and feline type C viruses. Cold Spring Harbor Symp Quant Biol 39:1159–1168

Van Beveren C, Van Straaten F, Galleshaw JA, Verma JM (1981) Nucleotide sequence of the genome of a murine sarcoma virus. Cell 27:97–108

Van de Ven WJM, Reynolds FH, Stephenson JR (1980a) The nonstructural components of polyproteins encoded by replication-defective mammalian transforming retroviruses are phosphorylated and have associated kinase activity. Virology 101:185–197

Van de Ven WJM, Reynolds FH, Nalewaik RP, Stephenson JR (1980b) Characterization of a 170,000 kd polyprotein encoded by the McDonough strain of feline sarcoma virus. J Virol 35:165–175

Velicer LF, Graves DC (1974) Properties of feline leukemia virus. II. In vitro labelling of the polypeptides. J Virol 14:700–703

Veronese F, Kelloff GJ, Reynolds FH, Hill RW, Stephenson JR (1982) Monoclonal antibodies specific to transforming polyproteins encoded by independent isolates of feline sarcoma virus. J Virol 43:896–904

Waneck GL, Rosenberg N (1981) Abelson leukemia virus induces lymphoid and erythroid colonies in infected fetal cell cultures. Cell 26:79–89

Weiss R, Teich N, Vamus H, Coffin J (1982) Molecular biology of tumor viruses, 2nd edn RNA tumor viruses. Cold Spring Harbor Laboratory, New York

Witte ON, Rosenberg N, Paskind M, Shields A, Baltimore D (1978) Identification of an Abelson murine leukemia virus-encoded protein present in transformed fibroblasts and lymphoid cells. Proc Natl Acad Sci USA 75:2488–2492

Witte ON, Dasgupta A, Baltimore D (1980) Abelson murine leukemia virus protein is phosphorylated in vitro to form phosphotyrosine. Nature 283:826–831

Wolfe LG, Deinhardt F (1978) Overview of viral oncology studies in *Saguinus* and *Callithrix* species. Primates Med 10:96–118

Wolfe LG, Deinhardt F, Theilen GH, Robin H, Kawakami T, Bustad LK (1971) Induction of tumors in marmoset monkeys by simian sarcoma virus type 1 (Lagothrix): A preliminary report. J Natl Cancer Inst 48:1115–1120

Wong-Staal F, Dalla-Favera R, Gelman EP, Manzari V, Szala S, Josephs SF, Gallo RC (1981) The v-*sis* gene of simian sarcoma virus is a new *onc* gene of primate origin. Nature 294:273–275

Ziegler SF, Whitlock CA, Goff SP, Gifford A, Witte ON (1981) Lethal effect of the Abelson murine leukemia virus transforming gene product. Cell 27:477–486

Asymmetric Chromosomal Distribution of Endogenous Retrovirus Loci in Chickens and Mice

ALLAN TEREBA

1	Introduction	29
2	Localization of Endogenous Chicken Retrovirus Loci	30
2.1	The Chicken Karyotype	31
2.2	*ev*1: A Commonly Occurring Retrovirus Locus	32
2.3	V^+-Associated Loci: *ev*2, *ev*7, and *ev*14	33
2.4	gs^-chf^--Associated Loci: *ev*4, *ev*5, *ev*8, and *ev*13	34
2.5	Transcribed but Structurally Defective Loci: *ev*3 and *ev*6	35
2.6	Why Clustering of *ev* Loci?	36
2.7	Localization of *ev* Loci in Diverse Chicken Breeds	37
3	Murine Endogenous Retrovirus Loci	37
3.1	Mouse Karyotype	38
3.2	Xenotropic Viral Loci	38
3.3	Ecotropic Viral Loci	40
3.3.1	Locally Clustered Loci	41
3.3.1.1	Chromosome 9	41
3.3.1.2	Chromosome 8	42
3.3.1.3	Chromosome 7	42
3.3.1.4	Chromosome 11	43
3.3.2	Unlinked Loci	43
3.4	Artificially Introduced Loci	44
3.5	Endogenous Mammary Tumor Virus Loci	44
4	Summary and Conclusions	45
References		46

1 Introduction

Retrovirus-like sequences have been found in the genomes of a large number of species (COFFIN 1982). These sequences are inherited in a Mendelian fashion, appear to be stable loci, and can be considered an integral part of an individual's genetic makeup. However, analysis of related species (FRISBY et al. 1979) and some related sublines (ROWE and KOZAK 1980) has revealed that these sequences are probably due to rare integration events following infections of gamete or embryonic tissue. The genetic loci resulting from this unique occurrence in animal virology provide an excellent opportunity to observe the interactions of exogenously introduced DNA with a vertebrate genome in a multitude of cell types and at different stages

St. Jude Children's Research Hospital, 332 North Lauderdale, Memphis, TN 38101, USA

Current Topics in Microbiology and Immunology, Vol. 107

of differentiation. Indeed, studies with artificially introduced endogenous retrovirus loci have indicated that the expression of these loci is controlled to some extent by the chromosomal location of the loci (JÄHNER and JAENISCH 1980; JAENISCH et al. 1981).

Many molecular analyses have been performed on endogenous retrovirus loci, especially in chickens and mice. As detailed in recent reviews (COFFIN 1982; ROVIGATTI and ASTRIN (to be published), several individual loci have been defined by characteristic viral-cell junction fragments obtained by restriction endonuclease digestions. This molecular characterization, along with detailed sequence determinations, has left the impression that these loci were and continue to be formed by an integration mechanism similar to that employed by current field isolates of retroviruses. Owing to the large number of distinct loci and apparent lack of physical linkage, this analysis also suggests that the integration process is random or at least not very selective.

During the last few years, the biochemical analysis of endogenous retrovirus loci has been supplemented by data concerning their chromosomal locations. Although the techniques used to obtain this chromosomal information can localize genes only to a relatively large chromosomal domain, they have produced some rather unexpected results in both the chicken and the murine systems. Contrary to previous predictions that the viral loci would be distributed randomly throughout the genome, it is becoming evident that at least a subgroup of these loci tend to be clustered in regional chromosomal domains. Whether this clustering is due to regional site-specific integration, selective pressures favoring retention of clusters, or a gene duplication mechanism remains to be determined. In any event an understanding of the mechanism surrounding the chromosomal organization and expression of these loci should provide greater insight into the evolution and regulation of a variety of cell genes.

2 Localization of Endogenous Chicken Retrovirus Loci

Extensive studies using restriction endonucleases have demonstrated the presence of 16 distinct retrovirus loci from a variety of inbred and outbred White Leghorn flocks (ASTRIN 1978; TEREBA and ASTRIN 1980; HUGHES et al. 1981; TEREBA et al. 1981; TEREBA and ASTRIN 1982). These loci have been designated *ev*1 to *ev*16 (ROVIGATTI and ASTRIN to be published) and most have been studied extensively with respect to associated phenotype (ASTRIN 1978; ASTRIN et al. 1979; ASTRIN et al. 1980; ASTRIN and ROBINSON 1979; CRITTENDEN and ASTRIN 1981; ROBINSON et al. 1979), genomic deletions (HAYWARD et al. 1980; HUGHES et al. 1981), and transcriptional activity (HAYWARD et al. 1980; BAKER et al. 1981). Studies on other breeds and commercial flocks have revealed the presence of many additional loci (ROVIGATTI and ASTRIN to be published; HUGHES et al. 1979b), but a systematic nomenclature has not been developed. Of the 16 defined loci in White Leghorns, nine have been localized to specific regions on macrochromosomes

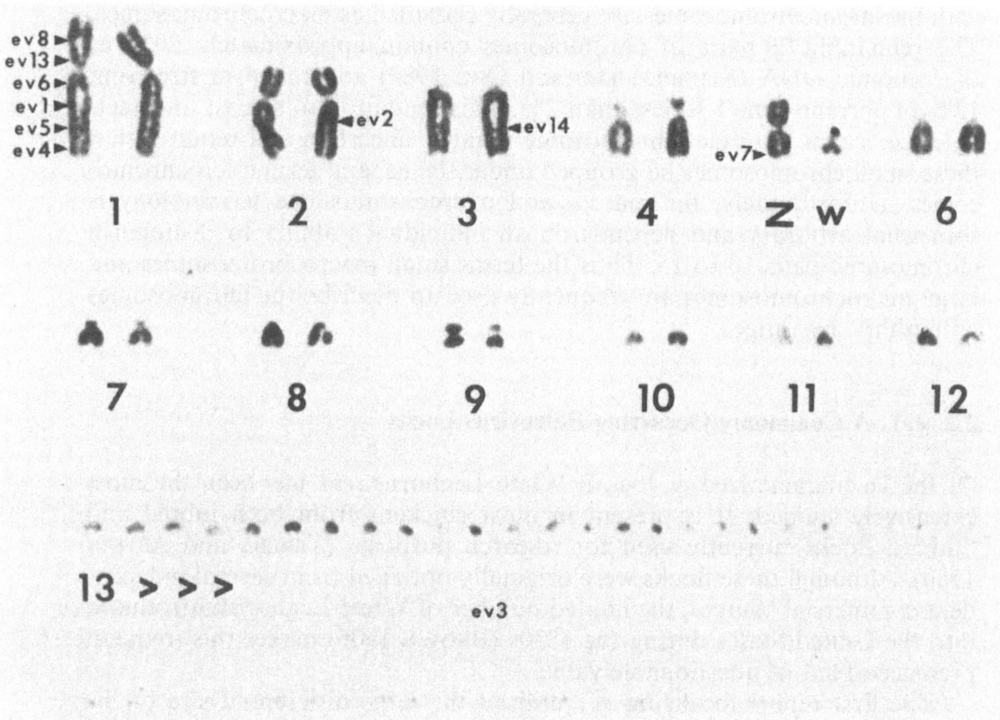

Fig. 1. Chromosomal location of chicken endogenous retrovirus loci. The chromosomal location of ten endogenous retrovirus loci in White Leghorn chickens was determined by in situ hybridization of a viral-specific probe to metaphase chromosomes from fibroblasts of chick embryos containing one or two viral loci as determined by restriction endonuclease digestion. Note that all loci containing 5′ terminal deletions (*ev*4, *ev*5, and *ev*6) and all gs^-chf^--associated loci (*ev*4, *ev*5, *ev*8, and *ev*13) are located on chromosome 1. (TEREBA 1981)

and a tenth locus has been assigned to one of the microchromosomes. To appreciate the nonrandom distribution of these loci and to help gain some understanding of the possible mechanisms responsible for this asymmetric distribution, the loci have been grouped into related phenotypes. In addition, this section is prefaced with a short description of the chicken karyotype to acquaint the reader with the advantages and disadvantages associated with avian chromosome analysis.

2.1 The Chicken Karyotype

The chicken genome consists of approximately 39 chromosome pairs, including the sex chromosomes Z and W (OHNO 1961). The first five autosomes and the Z chromosome contain approximately 65% of the genomic DNA (STUBBLEFIELD and ORO 1982) and can be easily recognized based on size and morphology (Fig. 1). Chromosome pairs 7 through 10 and the sex chromosome W can be distinguished in photomicrographs and along with the larger chromosomes are generally classified as macrochromosomes. The remaining 29 pairs of chromosomes contain approximately 20% of the genomic DNA (STUBBLEFIELD and ORO 1982) and range in size from 12% of chromosome 1 to less than 2%. This gradation to almost undetectable size leaves the exact chromosome number uncertain and requires that these small chromosomes be grouped under the general term microchromosomes. Unfortunately, the macro- and microchromosome terminology is somewhat arbitrary and depends on an individual's ability to distinguish chromosome pairs 10 to 15. Thus the terms small macrochromosomes and large microchromosomes are frequently used to describe the chromosomes within this size range.

2.2 *ev*1: A Commonly Occurring Retrovirus Locus

Of the 16 characterized *ev* loci in White Leghorns, *ev*1 has been the most extensively studied. It is present in most chickens from both inbred and outbred flocks currently used for research purposes (TEREBA and ASTRIN 1980). Although these flocks were originally obtained from several independent commercial sources, the limited number of White Leghorns introduced into the United States during the 1830s (BROWN 1906) makes this frequent presence of *ev*1 of questionable value.

The first report localizing *ev*1 utilized the large differential size of the chicken chromosomes (PADGETT et al. 1977). Metaphase chromosomes from the cell line MSB-1 were separated on a sucrose gradient; the DNA was extracted from each fraction and electrophoresed after digestion with a restriction endonuclease. After blotting onto a nitrocellulose filter, hybridization to viral-specific sequences was observed only in the fractions containing the largest chromosomes. This technique has since been refined by using the sucrose-fractionated chromosomes as a source for separating individual chromosomes based on DNA content using a fluorescent-activated cell sorter (STUBBLEFIELD and ORO 1982). These studies demonstrated the presence of *ev*1 on chromosome 1, the largest chicken chromosome.

Using a different approach, TEREBA and ASTRIN (1980) were able to localize *ev*1 to a defined region of chromosome 1. The DNA from fibroblasts of White Leghorn chick embryos was investigated for the presence of *ev* loci by restriction endonuclease analysis. Metaphase chromosomes from an embryo containing only *ev*1 were prepared and then hybridized in situ. The probe consisted of viral RNA hybridized via its poly A to high molecular weight ^{125}I-labeled sea urchin DNA tailed with poly BrUdR (TEREBA et al. 1979). Hybridization occurred specifically near the middle of the long arm of chromosome 1 (Fig. 1). This approach has since been utilized to detect nine other viral loci (see below) and in the process has confirmed the location of *ev*1 in 14 separate experiments using nine different embryos with a standard deviation of less than 2% of the chromosome length (TEREBA 1981; TEREBA et al. 1981; TEREBA and ASTRIN 1982; TEREBA and LAI 1982). Owing to this reproducibility and the concordance of the two experimental approaches, the chromosomal location of *ev*1 appears to be well established.

2.3 V^+-Associated Loci: *ev*2, *ev*7, and *ev*14

Of the 16 characterized *ev* loci, six are associated with the phenotype of virion production (HAYWARD et al. 1980; BAKER et al. 1981). In general, they are transcribed at low levels and all but *ev*7 produce infectious virions indicating that the complete viral genome is present and functional (ROVIGATTI and ASTRIN to be published). From a comparative standpoint, these loci are most like the exogenous viral loci resulting from infection of somatic tissue and would be the most likely source of new loci, assuming that the mechanism involved infection of gamete or zygote cells.

Of the six V^+-associated loci, three have been localized to specific regions of three distinct macrochromosomes by in situ hybridization (TEREBA et al. 1981). Although the technical approach was identical to the chromosomal localization of *ev*1, each embryo utilized had *ev*1 in addition to the viral locus of interest. Also, the use of Rous sarcoma virus RNA in the complex probe instead of a leukosis virus RNA resulted in hybridization of three different loci, two viral loci and the cellular c-*src* gene, in each experiment. In examining metaphase chromosomes from each embryo, hybridization over chromosome 1 and the chromosome group 10–12 was observed in all cases. In addition, the chromosome spreads that had *ev*2, *ev*7, and *ev*14 had an additional region of hybridization over chromosomes 2, Z, and 3, respectively. The hybridization to chromosome 1 and the chromosome group 10–12 was interpreted as hybridization to *ev*1 and the c-*src* gene which had previously been localized to a small macrochromosome by in situ hybridization (TEREBA et al. 1979; TEREBA and ASTRIN 1980) and by chromosome fractionation studies (PADGETT et al. 1977; HUGHES et al. 1979a). Thus by the process of elimination, *ev*2, *ev*7, and *ev*14 were assumed to be on chromosomes 2, Z, and 3, respectively. Analysis of the grain distribution along these chromosomes localized the individual loci to specific regions on each of the chromosomes as indicated in Fig. 1.

In addition to the localization of these viral loci by in situ hybridization, *ev*7 was localized independently using animal genetics (SMITH and CRITTENDEN 1981). In chickens, the males have two Z sex chromosomes while the females have a Z and a W chromosome. Thus, when examining a homozygous locus on the Z chromosome which is present in one line but absent in another, the results of a mating will depend on the line from which the female has come. SMITH and CRITTENDEN (1981) utilized line 15_B, which is homozygous for *ev*7, and lines 6_3 and 7_2, which lack *ev*7. When 15_B males (ZZ) were used in the matings, all 13 progeny were *ev*7-positive. However, when 15_B females were mated with line 6_3 or 7_2 males, only the male progeny contained the *ev*7 locus. All 14 female progeny from this set of crosses (which obtained their Z chromosome from the line 6_3 of 7_2 male) were *ev*7-negative. This set of experiments, while not providing as detailed a localization as the in situ hybridization experiments, did provide important confirmation of the reliability of the in situ hybridization technique.

2.4 gs^-chf^--Associated Loci: *ev*4, *ev*5, *ev*8, and *ev*13

Of the 16 characterized White Leghorn-associated viral loci, six appear to lack any transcriptional activity (HAYWARD et al. 1980; BAKER et al. 1981; ROVIGATTI and ASTRIN to be published). This is due in large part to deletions which eliminate either the 5′ promotor region (HAYWARD et al. 1980) or the structural genes (HUGHES et al. 1981). As shown above, the diverse chromosomal locations of the V^+-associated loci and *ev*1 are consistent with the theory that endogenous retrovirus loci are generated by exogenous infections and that the integration of these sequences is random or at least not very selective. The chromosomal localization of four of the six known gs^-chf^--associated loci casts some serious doubts on this theory as all four have been localized to chromosome 1.

The chromosomal localization of *ev*4, *ev*5, and *ev*8 utilized the same in situ hybridization technique described previously and relied on restriction endonuclease analysis to select embryos which contained only *ev*1 and one of the other three loci (TEREBA and ASTRIN 1982). The presence of the common *ev*1 locus did present some complications which were partly overcome by utilizing the inherent hybridization properties of the probe. Since the radioactive component of the probe was not a part of the hybridized sequences, any locus that did hybridize would contain approximately the same amount of radioactivity regardless of the size of the hybridized sequence. Using a probe that contained viral and *src* sequences, hybridization occurred on chromosome 1 and the chromosome group 10–12 in approximately a 2 to 1 ratio when examining embryos that contained *ev*1 and either *ev*4, *ev*5, or *ev*8. Since the chromosome group 10–12 has the c-*src* sequences, all four of the viral sequences were assumed to be on chromosome 1.

The data were strengthened by localizing the viral sequences at specific locations on chromosome 1. Since [^{125}I] emissions produce a majority of grains within a tenth to a fifth of the length of moderately spread number one chromosomes (ADA et al. 1966; ERTL et al. 1970), two peaks containing an equal number of grains would be expected in the above experiments unless the loci were close together. This two-peak result was obtained by TEREBA and ASTRIN (1982) in each case, and thus the chromosomal positions of *ev*4, *ev*5, and *ev*8 were determined (Fig. 1).

Confirmation of these chromosomal locations by animal genetics is fragmentary. Initial studies indicated that none of the *ev* loci were genetically linked (ASTRIN 1978). However, a combination of subsequent studies in which the segregation of *ev*4 and *ev*5 could be examined revealed seven recombinants in 35 offspring (ASTRIN 1978; ASTRIN et al. 1979; ASTRIN and ROBINSON 1979). This suggests a map distance of 20 units with a 95% confidence limit between 7 and 33 map units. Although correlation of map distances with chromosomal lengths is crude at best, these values agree quite well with the in situ hybridization data if one assumes that chromosome 1 has approximately 220 map units (SOMES 1979) and that the distance between *ev*4 and *ev*5 is 13% of chromosome 1 (Fig. 1). Although the genetic experiments do not place *ev*4 and *ev*5 on chromosome 1, the good agreement concerning the linkage of these two loci using in situ hybridization and animal genetics strongly suggests that the in situ hybridization data are valid.

A fourth viral locus associated with the gs^-chf^- phenotype has been located but lacks restriction endonuclease characterization. While developing their in situ hybridization technique, TEREBA et al. (1979) analyzed the metaphase chromosomes of a gs^-chf^- chick embryo. Three distinct peaks of silver grains were observed over chromosome 1. Subsequent chromosomal localization analysis of restriction endonuclease-characterized embryos from the same flock revealed that two of the peaks corresponded to the locations of *ev*1 and *ev*5 (TEREBA and ASTRIN 1982). The third peak located near the centromere (Fig. 1) had not been seen previously and thus was designated *ev*13. Whether this locus corresponds to a new viral locus not observed by restriction endonuclease analysis or is in fact *ev*15 or *ev*16, as defined by restriction analysis, must await determination of the chromosomal localization of these two loci.

2.5 Transcribed but Structurally Defective Loci: *ev*3 and *ev*6

Assuming a random integration mechanism, the localization of five of eight viral loci on chromosome 1 is somewhat disconcerting. This is beyond the limits of reasonable chance taking into account the relative size of chromosome 1. Even more odd is the fact that all four of the gs^-chf^--associated loci were located on this same chromosome. Assuming that there must be some structural basis underlying this apparent phenotypic clustering, TEREBA (1981) localized two additional loci, *ev*6 and *ev*3. The *ev*6 locus is transcribed but lacks the 5′ terminal LTR and the *gag* gene (HAYWARD et al. 1980). Structurally, this locus is very similar to *ev*4 and *ev*5, which also have 5′ terminal deletions (HAYWARD et al. 1980; HUGHES et al. 1981a). HAYWARD et al. (1980) have hypothesized that the transcriptional activity of this locus is due to the joining of an active cell promotor to this defective viral locus. If 5′ terminal deletions were a frequent occurrence in maintaining or generating these loci on chromosome 1, then *ev*6 should be located on chromosome 1 even though it is transcribed. The *ev*3 locus, on the other hand, has a 1.5 kb internal deletion spanning the *gag-pol* junction. While transcription occurs, no virions are produced (HAYWARD et al. 1980). Localization of this *ev*3 locus would suggest whether or not viral loci on chromosome 1 were associated with deletions in general.

Using in situ hybridization with metaphase chromosomes from a chick embryo containing *ev*1 and *ev*6, two peaks of grains were observed on chromosome 1. On peak corresponded to the location of *ev*1 while the other, presumably indicating *ev*6, was on the long arm near the centromere (Fig. 1).

Thus, all three loci known to contain 5′ terminal deletions are not only on chromosome 1 but also on the same arm. This clustered arrangement suggests that this type of deletion may be involved in the generation or maintenance of these loci.

In contrast, when metaphase chromosomes were analyzed from a line 6_3 chick embryo which contained *ev*1 and *ev*3, specific in situ hybridization occurred over chromosome 1 at the location of *ev*1 and over the microchromosome group. Thus, *ev*3 was assumed to be on one of the microchromosomes, and it was concluded that large deletions per se were not necessarily restricted to viral loci on chromosome 1.

2.6 Why Clustering of *ev* Loci?

While the chromosomal localization of *ev* loci in White Leghorns is far from complete, enough information has been obtained from specific flocks to suggest that an asymmetric distribution does exist. But is this clustering a statistical fluke? The localization of six of ten loci to chromosome 1 is statistically improbable ($P = 0.0038$) for a random distribution. Considering the limited number of loci analyzed though, this is not overly dramatic. However, taking into account the fact that all three loci containing 5′ terminal deletions are on chromosome 1, as are all four gs^-chf^--associated loci, it appears that this clustering has a physical basis and that the asymmetry is due to this subset of loci.

There are several possible explanations for the clustering of this subgroup of viral loci, but each theory has certain deficiencies. Site-directed integration is one possibility. Transposons, with which retroviruses appear to have many similarities (HISHINUMA et al. 1981; HUGHES et al. 1978; JU and SKALKA 1980; TEMIN 1980), are known to have preferred sites of integration. However, all the known endogenous chicken retrovirus loci examined are known to be highly conserved (COFFIN 1982) and it would be difficult to explain why V^+-associated loci are scattered throughout the genome while the gs^-chf^--associated loci are clustered.

A second source of clustering could be breeding patterns. As a result of (a) the very limited stock that initiated the White Leghorn flocks in the United States (BROWN 1906) and (b) selective breeding for particular traits (BACON 1979), it is possible that an artificial clustering occurred. However, traits of interest to breeders do not appear to be clustered on chromosome 1 (SOMES 1979). In addition, it seems improbable for an *ev*-ladened chromosome 1 to be selectively retained in a cross between *ev*-rich and *ev*-poor chickens. As chromosome 1 has at least 220 map units (SOMES 1979), all but adjacent viral loci would segregate independently if located as shown in Fig. 1.

A third hypothesis explaining clustering is that at least some of the loci on chromosome 1 were generated by gene duplication either at the locus level or through partial chromosome duplication. This model provides a mechanism to explain the common phenotype and suggests that 5′ terminal deletions are a frequent result of duplication. The presence of all six chromosome 1-associated loci in SPAFAS chickens would also be consistent with this hypothesis. However, owing to the spacing of the loci, certain restrictions must be imposed. Both the in situ hybridization data and animal genetic analysis show the viral loci to be spaced in a surprisingly uniform manner along the length of the chromosome. This eliminates the simple model of tandem gene duplication unless some mechanism for intrachromosomal dispersion was also involved. In this regard, duplication of part of chromosome 1 containing a progenitor *ev* locus appears to be the most likely event, but to propose any detailed mechanism would be pure speculation.

2.7 Localization of *ev* Loci in Diverse Chicken Breeds

Owing to the extensive use of White Leghorns in retrovirus research and the development of inbred lines from this breed, very little retrovirus research has been done with other breeds. A few studies using restriction endonuclease analysis on diverse breeds have shown a great deal of new *ev* loci (HUGHES et al. 1979b; ROVIGATTI and ASTRIN to be published). However, no genetic analysis of *ev* loci in these breeds has been attempted and only preliminary in situ hybridization analysis has been performed. In examining a series of White Plymouth Rock embryos from a single commercial flock, four distinct loci containing *ev*-related sequences were identified (PRIDGEN and TEREBA, unpublished observation). One locus was localized to the Z chromosome near the middle of the long arm; this was near but distinct from *ev7*. The other three loci were located on microchromosomes. Whether these loci represent another cluster set will have to await further analysis. In any event, no loci were detected on chromosome 1, suggesting that chickens in general do not have a propensity for chromosome 1-associated *ev* loci.

3 Murine Endogenous Retrovirus Loci

Contrary to the relatively low number of White Leghorn endogenous retrovirus loci, all strains of mice contain a complex maze of endogenous retrovirus loci. Each mouse contains 30–50 distinct loci complementary, at least in part, to murine leukemia viruses (STEFFEN and WEINBERG 1978; CANAANI and AARONSON 1979). Most of these loci contain sequences complementary to xenotropic viruses and for the most part are not expressed (LEVY 1978). However, several laboratory strains derived from Japanese breeds contain sequences complementary to ecotropic viruses which are frequently expressed or can be induced by a variety of agents (PINCUS 1980). Owing to the ability of ecotropic viruses to grow in mouse cells, it is not surprising that sequential analysis of mouse lines has revealed a gradual increase of ecotropic-associated viral loci presumably arising from infection of embryos or gonadal tissue (ROWE and KOZAK 1980).

In addition to these C-type retrovirus sequences, all strains of mice have about 500 loci containing sequences found in A-type particles (COFFIN 1982), and some 30 loci associated with a retrovirus-like 30S RNA species termed VL30 (BESMER et al. 1979; SCOLNICK et al. 1979). Many strains also possess a few loci related to B-type mammary tumor viruses (MORRIS et al. 1977). Although little has been done toward localizing VL30 and A-type loci, the manageable number of endogenous mammary tumor virus loci identified so far (17) and the variety of phenotypes observed (COFFIN 1982) make this a very useful system which has recently received considerable attention concerning chromosomal localization.

3.1 Mouse Karyotype

The mouse contains 19 pairs of autosomal chromosomes plus the sex chromosomes X and Y. All chromosomes are telocentric and vary in size by less than three fold (Fig. 2). This makes it impossible to distinguish individual chromosomes unless they are banded. For this reason, in situ hybridization has not been a very useful technique for examining the chromosomal location of defined loci. However, a great deal of genetic analysis has been performed and correlated with defined chromosomes through various other methods. Thus, there now exist a large number of genetic markers to correlate the murine retrovirus loci with specific locations on the various chromosomes. This approach, while useful, has a disadvantage in that possible regulatory genes may be mistaken for structural genes and loci near chromosomal markers will be localized first and thus may provide a biased perspective of the distribution of these viral loci.

3.2 Xenotropic Viral Loci

All mice so far examined have 30–50 loci containing sequences related to murine leukemia virus (PINCUS 1980; COFFIN 1982). Based on differences primarily in the *env* gene, these loci can be separated into two major groups, ecotropic and xenotropic (for reviews see LEVY 1978; PINCUS 1980; COFFIN 1982). The majority, and in some strains all, of these loci are related to the xenotropic class. This large xenotropic class, however, can be divided into two groups, the inducible class II viruses and the noninducible class III viruses (STEPHENSON et al. 1974; BARBACID et al. 1978). Based on sequence homologies with ecotropic viruses in the *gag* region and the fact that class II viruses are found in strains containing ecotropic viruses, it has been suggested that the class II viruses are recombinants of ecotropic and xenotropic viruses (BARBACID et al. 1978). However, restriction maps of several xenotropic viruses have suggested that this two-class division is oversimplistic and that recombinant viruses probably span both classes, resulting in more

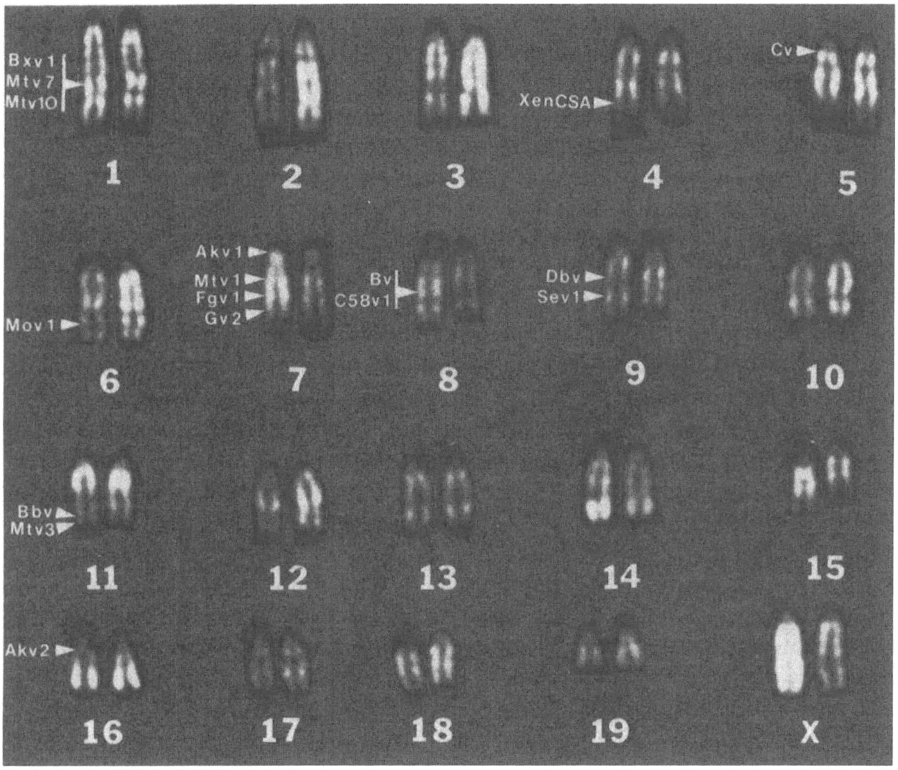

Fig. 2. Chromosomal location of 17 murine endogenous retrovirus loci. The approximate chromosomal location of endogenous mammary tumor virus loci (*Mtv*), xenotropic-associated C type loci (*Bxv*-1 and *XenCSA*), artificially introduced Moloney MuLV (*Mov*-1), and ecotropic-associated C type loci (remainder of loci) are designated on Q banded mouse chromosomes (courtesy of H. COON, H. COON, and M. SKLAR). The chromosomal location is based on genetic linkage data (presented in the text) and is therefore only approximate

of a gradation of genotypes (CHATTOPADHYAY et al. 1981). Thus, this xenotropic group of virus loci, while being relatively homogeneous on a gross level, is in fact quite heterogeneous.

Owing to the numerous cross-reacting xenotropic loci and the necessity to band chromosomes for detailed analysis, successful in situ hybridizations have not been performed. However, some analysis using restriction endonuclease technology and mouse–hamster somatic cell hybrids has been performed to look at the overall distribution of MuLV-related sequences (predominantly xenotropic) among mouse chromosomes (JOLICOEUR et al. 1980). Although the experiments were complicated by the heterogeneity in the karyotype of the hybrid clones, marked deviations from a random distribution were observed in the chromosomal location of these viral loci. In particular, chromosome 4 appeared to be rich in viral loci while chromosomes 5 and 15 were underrepresented. As a confirmation that chromosome 4 contained several viral loci, restriction endonuclease analysis of a hybrid that contained only chromosome 4 revealed several viral-specific bands. In a more refined analysis using a somatic cell hybrid containing the distal portion of chromosome 1, three xenotropic viral loci were identified (C. KOZAK, personal communication). These experiments demonstrate the utility of employing restriction endonuclease analysis with somatic cell hybrids to localize this large group of xenotropic loci. However, owing to the limited number of loci analyzed in detail so far, it is impossible to say whether the nonrandomness found in the overall distribution is due to statistical variations or to real domains of clustered loci.

The localization of xenotropic loci to distinct sites on a particular chromosome is very limited and has relied on genetic mapping of the few loci that are expressed. BALB/c mice contain an IdU-inducible locus which follows simple Mendelian genetic patterns when crossed with expression-negative strains (AARONSON and STEPHENSON 1973). This locus, termed *Bxv*-1, was mapped by KOZAK and ROWE (1978) to chromosome 1. Additional studies by KOZAK and ROWE (1980a) using animal genetics and somatic cell hybrids refined the location of this locus in five strains. The map position was determined as *Pep*-3–19.5–*Bxv*-1–5–*Lp* and its estimated location is shown in Fig. 2. As described below, this is very close to the location of two mammary tumor virus loci. Since virus induction was being examined and not viral sequences, the possibility remains that this locus represents a regulatory gene. However, somatic cell hybrids linking this phenotype only to chromosome 1, as well as analogies from the chicken *ev* loci, strongly suggest that this locus does represent one of the xenotropic-associated viral genes.

A second locus, associated with high levels of xenotropic gp70 antigen on the cell surface of thymocytes (*XenCSA*), has also been mapped by animal genetics (MORSE et al. 1979). This locus is closely linked (4 map units) to the *Fv*-1 restriction locus on chromosome 4 and its approximate position is shown in Fig. 2. Although it has been suggested that this locus is regulatory in nature (MORSE et al. 1979), the analogy with *Bxv*-1 and the avian *ev* loci, which are independently controlled (ROVIGATTI and ASTRIN 1983), suggests that this locus represents another structural virus locus.

3.3 Ecotropic Viral Loci

Unlike the xenotropic-associated loci, ecotropic viral loci are found in only a subset of mouse strains (PINCUS 1980), with most positive strains containing from one to three loci (JENKINS et al. 1982). In addition, a restriction endonuclease analysis of ecotropic-specific sequences in 53 inbred strains of mice revealed a total of only 14 distinct loci (JENKINS et al 1982), several of which have already been mapped by animal genetics. Although many more loci are undoubtedly present in wild mice and, as discussed below, new loci are continually being generated, the currently characterized ecotropic loci provide a sufficient number to begin assessing the organization of these loci in the mouse genome.

Owing to the frequent expression of these viral loci and the ability to cross viral-positive with viral-negative strains, animal genetics has proven very useful in localizing this class of loci. Currently, loci have been localized to seven different chromosomes containing approximately 35% of the mouse genome. This is very close to the prediction for a random distribution and is consistent with what was found with the V^+-associated loci in White Leghorn chickens. However, the local clustering of several of these loci with each other and with mammary tumor virus loci opens up the possibility that certain chromosomal locations are preferred either as integration sites or as being most suitable for disrupting the mouse genome without causing deleterious effects.

3.3.1 Locally Clustered Loci

3.3.1.1 Chromosome 9

The clustering of ecotropic viral loci can be demonstrated on several chromosomes. In several cases these clusters of viral loci contain unrelated genomes (i.e., B-type and C-type) or ecotropic viral loci present in different mouse strains. However, the *Dbv–Sev*-1 cluster provides a potentially interesting situation. DBA and P/J mice contain a highly expressed ecotropic viral locus designated *Dbv* or *emv*-3. JENKINS et al. (1981) used animal genetics to show that this viral locus was closely associated with, if not within, the dilute *d* locus on chromosome 9 (Fig. 2). The fact that a spontaneous revertant at the *d* locus lacked the *Dbv* viral sequences suggests that this viral genome integrated within this gene. SEA/GnJ mice are derived from a cross between P/J mice and BALB/c mice, the latter strain apparently containing only the Cv (*emv*-1) ecotropic-associated viral locus (JENKINS et al. 1982). Restriction endonuclease mapping suggests that this hybrid strain contains both the *Cv* poorly inducible locus and the *Dbv* locus. However, when animal genetics were performed with the SEA/Gn mice, an inducible viral locus, *Sev*-1, was observed on chromosome 9 but appeared distinct from *Dbv* (KOZAK and ROWE 1982). Three factor crosses gave the map positions as *Mpi*-15–15–*Mod*-1–20–*Sev*-1 (see Fig. 2 for approximate chromosomal position).

Using the same markers for *Dbv*, the map positions are *Mpi*-1–11–*Dbv*–9–*Mod*-1. Thus, *Dbv* and *Sev*-1 appear to be separated by approximately 29 map units. Several explanations for this discrepancy are possible. First, the location of *Sev*-1 may be in error and *Dbv* and *Sev*-1 are really allelic. Although possible, this is unlikely as two separate mating experiments totaling 141 progeny would have to be in error. Second, a rare integration event could have occurred, resulting in the addition of a new locus, *Sev*-1, from the parental *Dbv* locus. This would have had to occur after the formation of the SEA/Gn hybrid mice because of the restriction endonuclease analysis (JENKINS et al. 1982) and it would be very unusual to have the new genome integrated so closely to the parental locus. Third, some mechanism could have transposed the *Dbv* locus with or without adjacent genes to the *Sev*-1 site during one of the matings to generate the SEA/Gn strain or the mice used to analyze the *Sev*-1 locus. In any case, the question of allelism of these loci can and should be examined.

3.3.1.2 Chromosome 8

C58 mice contain several ecotropic V loci (STEPHENSON and AARONSON 1973). Two of these loci have been bred into an NFS background and one high expressing locus, *C58v*-1, localized by animal genetics (KOZAK and ROWE 1982). The locus was localized to chromosome 8 with a map order of *C58v*-1–26–*Es*-1–21–*Gr*-1 (Fig. 2): The *C58v*-1 locus was also mapped at 4 units form the E^{SO} gene but gene order was not determined.

The C57BL/10 mouse carries a poorly expressed ecotropic locus which has been assigned by restriction endonuclease analysis as *emv*-2 (JENKINS et al. 1982). When this locus, also designated as *Bv,* was analyzed by animal genetics, its location was assigned to chromosome 8 with a map position of *Gr*-1–21–*Es*-1–26–*Bv* (KOZAK and ROWE 1982). This put the *Bv* locus very close to, if not allelic with, the *C58v*-1 locus (Fig. 2). However, because of the marked difference in phenotypic expression, these two loci were assumed to be distinct. Owing to the common ancestry of the C57 and C58 strains (POTTER and KLEIN 1979), it is possible that some mutation event occurred after separation of these strains, altering the phenotype of one of the viral alleles. In any case, it becomes very important to determine whether these two viral loci are allelic in order to explain the mechanism for their close proximity.

3.3.1.3 Chromosome 7

Although chromosome 7 contains only 5%–6% of the mouse genome, it contains 3 of the 11 ecotropic viral loci mapped to date. While this is not a very rare event for a random distribution ($P = 0.025$), it is somewhat unusual and deserves some attention. AKR mice contain two loci associated with ecotropic virus production (ROWE 1972). One of these loci, *Akv*-1 (*emv*-11), was localized through animal genetics to chromosome 7, 15 map units proximal to *Gpi*-1 (ROWE et al. 1972). The approximate chromosomal position is shown in Fig. 2. Subsequent studies employing reassociation kinetics (CHATTOPADHYAY et al. 1975) and T1 fingerprinting (ROMMELAERE et al. 1977) have convincingly shown this to be a structural viral genome.

C3H/FgLw mice contain multiple V loci. Two of these loci, designated *Fgv*-1 and *Fgv*-2, were bred into NIH mice and subsequently into NFS mice for genetic analysis. The *Fgv*-1 locus was assigned to chromosome 7 owing to its linkage with the *C* (color) gene. Subsequent three-point cross-analysis showed the gene map to be *C*–8–*Hbb*–3–*Fgv*-1 (KOZAK and ROWE 1982). This is approximately 55 map units from *Akv*-1 (Fig. 2). The *Fgv*-2 locus has not been mapped yet.

Several strains of mice contain an antigen on their thymocyte membrane that is related to ecotropic-associated glycoprotein (gp70) of murine leukemia virus. Expression of this antigen, designated G_{IX}, appears dependent on the inheritance of two independently segregating genes, *Gv*-1 and *Gv*-2 (STOCKERT et al. 1971; OLD and STOCKERT 1977). Although there is no direct evidence linking these loci with structural genes, the evidence from the chicken system would strongly suggest that this is the case. Using animal genetics, the *Gv*-2 locus was localized to chromosome 7 near the distal end in the 129 strain of mice (H.C. MORSE III, personal communication). Its approximate chromosomal position is shown in Fig. 2. Thus, the three ecotropic-associated viral loci are distributed along the entire length of chromosome 7 and do not appear to be the result of a small region favored for integration and/or maintenance of viral loci. Interestingly, as discussed below, one of the four localized mammary tumor virus loci is also on this chromosome (Fig. 2).

3.3.1.4 Chromosome 11

The mouse line B10.BR contains a locus, *Bbv*, that is associated with high levels of B tropic ecotropic leukemia virus. This locus probably originates from the germ line insertion of a B tropic viral genome that is typically generated by recombination events in older *Fv*-1^b mice. In initial crosses, this locus segregated with the *Es*-3 locus on chromosome 11 (KOZAK and ROWE 1982). A second cross employing the *Re* locus gave the map position *Re*–16–*Bbv*–12–*Es*-3 (Fig. 2). As discussed below, the mammary tumor virus locus *Mtv*-3 has also been localized to chromosome 11 approximately 4 map units from *Es*-3. Thus, *Bbv* and *Mtv*-3 are between 8 and 16 map units apart.

3.3.2 Unlinked Loci

Although several of the ecotropic-associated viral loci show linkage to each other, a couple of these loci appear isolated. As mentioned above, AKR mice contain two loci that appear responsible for virus production and one has been localized to chromosome 7. The other locus, *Akv*-2 (*emv*-12), was localized by a combination of somatic cell hybrids and standard breeding techniques (KOZAK and ROWE 1980b). The *Akv*-2 locus was first placed into an NIH background and then hamster/mouse cells were made from cells of this cross. This allowed the localization of *Akv*-2 to chromosome 16. Animal crosses subsequently positioned *Akv*-2 five map units from the *md* locus. This is near the centromere (Fig. 2). The use of somatic cell hybrids along with T1 fingerprinting analysis (ROMMELAERE et al. 1977) showed that this locus was probably the site of a structural viral genome.

In contrast to high virus expression lines, several strains of mice show a sporadic expression relatively late in life and in low titers. The locus responsible for this expression, *Cv* (*emv*-1), was localized in a variety of mouse strains, including BALB/c mice (KOZAK and ROWE 1979; IHLE et al. 1979; KOZAK and ROWE 1982). Using BALB/c–hamster somatic cell hybrids, the *Cv* locus was identified on chromosome 5. Subsequent genetic analysis revealed the map position as *Cv–Pgm-1–Gus*. A more detailed analysis using A/J mice (KOZAK and ROWE 1982) placed the *Cv* locus very close to the *Hx* locus (0/17 recombinants), and the approximate location is shown in Fig. 2.

3.4 Artificially Introduced Loci

Certain ecotropic V^+ strains, such as AKR mice, have a propensity to accumulate new endogenous viral loci during years of breeding (ROWE and KOZAK 1980). This addition of viral loci is apparently due to infection of oocytes or the early developing embryo, as only V^+ females have given rise to offspring with additional endogenous loci. Although these new loci have not been mapped, they are not linked to the parental locus. In trying to mimic this situation, Jaenisch and colleagues (JAENISCH 1980) infected 4–8 cell embryos of BALB/c mice with Mo-MLV. Several of the embryos that developed into mice contained endogenous-like viral loci. In general, these new loci were not linked, although only one, designed *Mov*-1, was localized to chromosome 6 using mouse–hamster somatic cell hybrids (BREINDL et al. 1979). Animal genetic analysis has confirmed this chromosomal association and has shown a weak linkage to the *Wa*-1 locus. A tentative location of *Mov*-1 is shown in Fig. 2.

3.5 Endogenous Mammary Tumor Virus Loci

Most inbred strains of mice and many wild mice contain DNA sequences related to murine mammary tumor viruses (MORRIS et al. 1977). Typically, inbred strains will contain from three to eight distinct loci (COHEN and VARMUS 1979; MICHALIDES et al. 1981; MORRIS et al. 1979; LONG et al. 1980) but so far an examination of 19 different strains has revealed only 17 distinct loci using a combination of restriction endonuclease technology and phenotypic expression (COFFIN 1982). Owing to this manageable number of loci, significant progress has been made in the chromosomal localization of these viral genomes.

In general, a genetic analysis of 11 viral loci employing restriction endonuclease technology showed independent segregation (TRAINA et al. 1981), suggesting that this class of loci is widely distributed among the mouse chromosomes. An exception to this was the relatively close linkage of *Mtv*-7 and *Mtv*-10 (12 map units). In addition, both of these viral loci showed close linkage to markers on the distal end of chromosome 1, *Mtv*-10–12–*Mls* and *Mtv*-7–4–*Eph*-1. Figure 2 shows the approximate location of these two viral loci and demonstrates that these loci are very close to the xenotropic locus *Bxv*-1. Thus, the distal end of chromosome 1 appears to be relatively enriched in viral loci.

Two other loci, *Mtv*-1 and *Mtv*-3, have been localized to distinct chromosomal positions. Using animal genetics, VAN NIE and VERSTRAETEN demonstrated an association between MMTV viral antigen and the *C* locus on chromosome 7 when analyzing $C3H_f$ and DBA_f mice (VAN NIE and VERSTRAETEN 1975; VERSTRAETEN and VAN NIE 1978). The recombination units were between 20 and 29. When TRAINA et al. (1981) analyzed MMTV loci, using a combination of restriction endonuclease analysis and animal genetics, they found a viral locus designated unit V on chromosome 7. Analysis of multiple markers revealed a map position of *Gpi*-1–*Tam*-1–unit V–*Mdr*–*Hbb*. This placed the viral locus somewhat closer to the *C* locus than the Van Nie and Verstraeten reports. However, it was concluded that because of the close proximity, unit V and *Mtv*-1 were probably the same locus. The tentative chromosomal position of *Mtv*-1 (as shown in Fig. 2) increases the viral load of chromosome 7.

The *Mtv*-3 locus, present in GR/Mtv-2$^-$ mice, was localized by animal genetics to chromosome 11 (NUSSE et al. 1980). In two sets of matings, the viral locus was closely associated with *Es*-3 (3% and 5% recombinants). As noted above, this places *Mtv*-3 very close to the *Bbv* ecotropic locus (Fig. 2).

Three additional *Mtv* loci have been identified with specific chromosomes although the locations of the loci on the chromosomes have not yet been defined. TRAINA et al., using animal genetics, were able to associate *Mtv*-14 with two loci associated with chromosome 6: *Lyt*-2, a lymphocyte alloantigen, and *GgC*, γ-glutamyl cyclotransferase. The linkage, though not extensive, was quite significant – 9/37 and 11/37 recombinants respectively, with a tentative gene order *Mtv*-14–*Lyt*-2–*GgC*. *Mtv*-2 was reported to be present on chromosome 18 linked to the *Tw* gene (NUSSE et al. 1980), although the degree of linkage was not presented. Finally, in analyzing mouse–hamster somatic cell hybrids by hybridization kinetics and restriction endonucleases (MORRIS et al. 1979), a single *Mtv* locus was observed in a hybrid containing only mouse chromosome 4. Since the restriction fragments do not match any of the known *Mtv* loci, this may be a new *Mtv* locus.

4 Summary and Conclusions

With the advent of modern recombinant DNA technology, entirely new areas of research have been opened up. It is now possible not only to examine the molecular events surrounding the rearrangements of genetic material in eukaryotic cells but also to actually alter their genetic composition by adding new genetic information in discrete well-defined packets. The examination of endogenous retrovirus loci is in essence an examination of nature's experimentation in this field, and an understanding of this natural event should help us not only to overcome the problems associated with artificially introducing genes into a whole vertebrate organism but also to understand the cellular gene organization that evolution has produced.

Cursory examination of endogenous retrovirus integration sites by restriction endonuclease technology has shown that exogenous DNA is inserted in a wide range of chromosomal sites with no apparent consensus integration sequence. Examination of the chromosomal location of these loci in chickens and mice, however, has shown quite a definite asymmetric distribution. In the White Leghorn chicken, six of ten viral loci localized so far have been identified on chromosome 1. This clustering appears to be the result of a subset of loci which lack active viral transcriptional promotors and in half the cases lack the entire 5′ long terminal repeat. This phenotypic and structural similarity, the evenly spaced distribution of viral loci along chromosome 1, and the apparent absence of viral clustering on chromosome 1 in White Plymouth Rock chickens all point to a chromosomal alteration which has resulted in the multiple duplication of a progenitor viral locus, most likely *ev*1. A detailed molecular analysis of these six loci and the chromosomal regions surrounding them may not only help confirm this mechanism but also provide some insights into the evolution of multigene families.

The overall chromosomal distribution of murine endogenous retrovirus loci remains unclear owing to the abundance of these loci. As with the chicken loci, restriction endonuclease analysis has shown multiple integration sites of all four major types of viral-like sequences. A cursory analysis at the chromosomal level suggests a random distribution of the ecotropic-associated loci. This is consistent with the disperse distribution of V^+-associated chicken endogenous viral loci and suggests that most of these loci are the result of independent integration events. However, within this gross random distribution, several clusters of two or more loci exist. One of these clusters could be explained by a mutational event that resulted in phenotypically distinct alleles associated with different strains of mice. However, the reason for frequent clustering of heterologous viral loci ramains unclear.

In any event, the further examination of the chromosomal distribution of these loci in mice and chickens should confirm the existence of subsets of viral loci that have been generated by mechanisms other than direct integration of exogenously introduced viral genomes. Since these viral loci are in essence cellular genes, the mechanisms responsible for their duplication, translocation, and subsequent maintenance in a diversity of cell types should mirror the mechanisms used by other cellular genes. Thus, by analyzing the viral loci, a greater understanding concerning the evolution of vertebrate gene organization should evolve.

References

Aaronson SA, Stephenson JR (1973) Independent segregation of loci for activation of biologically distinguishable RNA C-type viruses in mouse cells. Proc Natl Acad Sci USA 70:2055–2058

Ada G, Humphrey JH, Askonas BA, McDevitt HO, Nossal GJV (1966) Correlation of grain counts with radioactivity (^{125}I and tritium) in autoradiography. Exp Cell Res 41:557–572

Astrin SM (1978) Endogenous viral genes of the White Leghorn chicken: common site of residence and sites associated with specific phenotypes of viral gene expression. Proc Natl Acad Sci, USA 75:5941–5945
Astrin SM, Robinson HL (1979) Gs, an allele of chickens for endogenous avian leukosis viral antigens, segregates with *ev*3, a genetic locus that contains structural genes for virus. J Virol 31:420–425
Astrin SM, Crittenden LB, Buss EG (1979) *Ev*3, a structural gene locus for endogenous virus, segregates with the $gs^{+}chf^{+}$ phenotype in matings of line 6_3 chickens. Virology 99: 1–9
Astrin SM, Crittenden LB, Buss EG (1980) *Ev*2, a genetic locus containing structural genes for endogenous virus, codes for Rous-associated virus type 0 produced by line 7_2 chickens. J Virol 33:250–255
Bacon LD (1979) Origin of inbred and genetically defined strains of chickens. In: Altman PL, Katz DD (eds) Inbred and genetically defined strains of laboratory animals. Federation of American Societies for Experimental Biology, Bethesda, p 607
Baker B, Robinson H, Varmus HE, Bishop JM (1981) Analysis of endogenous avian retrovirus DNA and RNA: viral and cellular determinants of retrovirus gene expression. Virology 114:8–22
Barbacid M, Robbins KC, Hino S, Aaronson SA (1978) Genetic recombination between mouse type C RNA viruses: a mechanism for endogenous viral gene amplication in mammalian cells. Proc Natl Acad Sci USA 75:923–927
Besmer P, Olshevsky U, Baltimore D, Dolberg D, Fan H (1979) Virus-like 30S RNA in mouse cells. J Virol 29:1168–1176
Breindl M, Doehmer J, Willecke K, Dausman J, Jaenisch R (1979) Germ line integration of Moloney leukemia virus: identification of the chromosomal integration site. Proc Natl Acad Sci USA 76:1938–1942
Brown E (1906) Races of domestic poultry. Arnold, London, p 40
Canaani E, Aaronson SA (1979) Restriction enzyme analysis of mouse cellular type C viral DNA: emergence of new viral sequences in spontaneous AKR/J lymphomas. Proc Natl Acad Sci USA 76:1677–1681
Chattopadhyay SK, Rowe WP, Teich NM, Lowy DR (1975) Definitive evidence that the murine C-type virus inducing locus AKV-1 is viral genetic material. Proc Natl Acad Sci USA 72:906–910
Chattopadhyay SK, Lander MR, Gupta S, Rands E, Lowy DR (1981) Origin of mink cytopathic focus-forming (MCF) viruses: comparison with ecotropic and xenotropic murine leukemia virus genomes. Virology 113:465–483
Coffin J (1982) Endogenous viruses. In: Weiss R, Teich N, Varmus H, Coffin J (eds). RNA tumor viruses. Cold Spring Harbor Laboratory, Cold Spring Harbor, pp 1109–1203
Cohen JC, Varmus HE (1979) Endogenous mammary tumor virus DNA varies among wild mice and segregates during inbreeding. Nature 278:418–423
Crittenden LB, Astrin SM (1981) Independent segregation of *ev*2 and *ev*10, genetic loci for spontaneous production of endogenous avian retroviruses. Virology 110:120–127
Ertl HH, Feinendegen LE, Heiniger HJ (1970) Iodine-125, a tracer in cell biology: physical properties and biological aspects. Phys Med Biol 15:447–456
Frisby DT, Weiss RA, Roussel M, Stehelin D (1979) The distribution of endogenous chicken retrovirus sequences in the DNA of galliform birds does not coincide with avian phytogenetic relationships. Cell 17:623–634
Hayward WS, Braverman SB, Astrin SM (1980) Transcriptional products and DNA structure of endogenous avian proviruses. Cold Spring Harbor Symp Quant Biol 44:1111–1122
Hishinuma F, DeBona PJ, Astrin SM, Skalka AM (1981) Nucleotide sequence of acceptor site and termini of integrated avian endogenous provirus *ev*1: integration creates a 6 bp repeat of host DNA. Cell 23:155–164
Hughes SH, Shank PR, Spector DH, Kung H-J, Bishop JM, Varmus HE, Vogt PK, Breitman ML (1978) Proviruses of avian sarcoma virus are terminally redundant co-extensive with unintegrated linear DNA, and integrate at many sites. Cell 15:1397–1410
Hughes SH, Stubblefield E, Payvar F, Engel JD, Dodgson JB, Spector D, Cordell B, Schimke RT, Varmus HE (1979a) Gene localization by chromosome fractionation: globin genes

are on at least two chromosomes and three estrogen-inducible genes are on three chromosomes. Proc Natl Acad Sci USA 76:1348–1352

Hughes SH, Farhang P, Spector D, Schimke RT, Robinson HL, Payne GS, Bishop JM, Varmus HE (1979b) Heterogeneity of genetic loci in chickens: analysis of endogenous viral and nonviral genes by cleavage of DNA with restriction endonucleases. Cell 18:347–359

Hughes SH, Bishop JM, Varmus HE (1981) Organization of the endogenous proviruses of chickens: implications for origin and expression. Virology 108:189–207

Ihle JN, Joseph DR, Domotor JJ Jr (1979) Genetic linkage of C3H/HJ and BALB/c endogenous ecotropic C-type viruses to phosphoglucomutase-1 on chromosome 5. Science 204:71–73

Jaenisch R (1980) Germ line integration and Mendelian transmission of exogenous type C viruses. In: Stephenson J (ed) Molecular biology of RNA tumor viruses. Academic, New York, pp 131–162

Jaenisch R, Jähner D, Nobis P, Simon I, Lohler J, Harbers K, Grotkopp G (1981) Chromosomal position and activation of retroviral genomes inserted into the germline of mice. Cell 24:519–529

Jähner D, Jaenisch R (1980) Integration of Moloney leukaemia virus into the germ line of mice: correlation between site of integration and virus activation. Nature 287:456–458

Jenkins NA, Copeland NG, Taylor BA, Lee BK (1981) Dilute(d) coat colour mutation of DBA/2J mice is associated with the site of integration of an ecotropic MuLV genome. Nature 293:370–374

Jenkins NA, Copeland N, Lee B, Taylor B (1982) In: Weiss R, Teich N, Varmus H, Coffin J (eds) RNA tumor viruses. Cold Spring Harbor Laboratory, Cold Spring Harbor, pp 1150–1151

Jolicoeur P, Rassant E, Kozak C, Ruddle F, Baltimore D (1980) Distribution of endogenous murine leukemia virus DNA sequences among mouse chromosomes. J Virol 33:1229–1235

Ju G, Skalka AM (1980) Nucleotide sequence analysis of the long terminal repeat (LTR) of avian retroviruses: structural similarities with transposable elements. Cell 22:379–386

Kozak CA, Rowe WP (1978) Genetic mapping of xenotropic leukemia virus-inducing loci in two mouse strains. Science 199:1448–1449

Kozak CA, Rowe WP (1979) Genetic mapping of the ecotropic murine leukemia virus-inducing locus of BALB/c mouse to chromosome 5. Science 204:69–71

Kozak CA, Rowe WP (1980a) Genetic mapping of xenotropic murine leukemia virus-inducing loci in five mouse strains. J Exp Med 152:219–228

Kozak CA, Rowe WP (1980b) Genetic mapping of the ecotropic virus-inducing locus AKV-2 of the AKR mouse. J Exp Med 152:1419–1423

Kozak CA, Rowe WP (1982) Genetic mapping of ecotropic murine leukemia virus-induced loci in six inbred strains. J Exp Med 155:524–534

Levy JA (1978) Xenotropic type C viruses. Curr Top Microbiol Immunol 79:109–213

Long CA, Dumaswala UJ, Tancin SL, Vaidya AB (1980) Organization and expression of endogenous murine mammary tumor virus genes in mice congenic at the H-2 complex. Virology 103:167–177

Michalides RE, Wagenaar E, Groner B, Hynes NE (1981) Mammary tumor virus proviral DNA in normal murine tissue and non-virally induced mammary tumors. J Virol 39:367–376

Morris VL, Medeiros E, Ringold GM, Bishop JM, Varmus HE (1977) Comparison of mouse mammary tumor virus-specific DNA in inbred, wild and Asian mice and in tumors and normal organs from inbred mice. J Mol Biol 114:73–91

Morris VL, Kozak C, Cohen JC, Shank PR, Jolicoeur P, Ruddle F, Varmus HE (1979) Endogenous mouse mammary tumor virus DNA is distributed among multiple mouse chromosomes. Virology 92:46–55

Morse HC III, Chused TM, Hartley JW, Mathieson BJ, Sharrow SO, Taylor BA (1979) Expression of xenotropic murine leukemia viruses as cell-surface gp70 in genetic crosses between strains DBA/2 and C57BL/6. J Exp Med 149:1183–1196

Nusse R, DeMoes J, Hilkens J, Van Nie R (1980) Localization of a gene for expression of mouse mammary tumor virus antigens in the GR/Mtv-2^- mouse strain. J Exp Med 152:712–719

Ohno S (1961) Sex chromosomes and microchromosomes of *Gallus domesticus*. Chromosoma 11:484–498

Old LJ, Stockert E (1977) Immunogenetics of cell surface antigens of mouse leukemia. Annu Rev Genet 11:127–160

Padgett TG, Stubblefield E, Varmus HE (1977) Chicken macrochromosomes contain an endogenous provirus and microchromosomes contain sequences related to the transforming gene of ASV. Cell 10:649–657

Pincus T (1980) The endogenous murine type C virus. In: Stephenson J (ed) Molecular biology of RNA tumor viruses: Academic, New York, pp 77–130

Potter M, Klein J (1979) Genealogy of the more commonly used inbred mouse strains. In: Altman PL, Katz DD (eds) Inbred and genetically defined strains of laboratory animals. Federation of American Societies for Experimental Biology, Bethesda, p 16

Robinson HL, Astrin SM, Salazar FH (1979) V-15_B, an allele of chickens for the production of a noninfectious avian leukosis virus. Virology 99:10–20

Rommelaere J, Faller DV, Hopkins N (1977) RNase T_1-resistant oligonucleotides of AKV-1 and AKV-2 type C viruses of AKR mice. J Virol 24:690–694

Rovigatti UG, Astrin SM (1983) Avian endogenous viral genes. Curr Top Microbiol Immunol 103:1–21

Rowe WP (1972) Studies of genetic transmission of murine leukemia viruses by AKR mice. I crosses with Fv-1^n strains of mice. J Exp Med 136:1272–1285

Rowe WP, Kozak CA (1980) Germ-line reinsertions of AKR murine leukemia virus genomes in AKV-1 congenic mice. Proc Natl Acad Sci USA 77:4871–4874

Rowe WP, Hartley JW, Bremmer T (1972) Genetic mapping of a murine leukemia virus-inducing locus of AKR mice. Science 178:860–862

Scolnick EM, Vass WC, Howk RS, Duesberg PH (1979) Defective retrovirus-like 30S RNA species of rat and mouse cells are infectious if packaged by C helper virus. J Virol 29:964–972

Smith EJ, Crittenden LB (1981) Segregation of chicken endogenous viral loci *ev*7 and *ev*12 with the expression of infectious subgroup E avian leukosis viruses. Virology 112:370–373

Somes RG Jr (1979) New linkage groups and revised chromosome map of the domestic fowl. J Hered 69:401–403

Steffen D, Weinberg RA (1978) The integrated genome of murine leukemia virus. Cell 15:1003–1010

Stephenson JR, Aaronson SA (1973) Segregation of loci for C-type virus induction in strains of mice with high and low incidence of leukemia. Science 180:865–866

Stephenson JR, Tronick SR, Reynolds RK, Aaronson SA (1974) Isolation and characterization of C-type viral gene products of virus-negative mouse cells. J Exp Med 139:427–438

Stockert E, Old LJ, Boyse EA (1971) The G_{IX} system. A cell surface allo-antigen associated with murine leukemia virus; implication regarding chromosomal integration of the viral genome. J Exp Med 133:1334–1355

Stubblefield E, Oro J (1982) The isolation of specific chicken macrochromosomes by zonal centrifugation and flow sorting. Cytometry 2:273–281

Temin HM (1980) Origin of retroviruses from cellular movable genetic elements. Cell 21:599–600

Tereba A (1981) 5′-Terminal deletions are a common feature of endogenous retrovirus loci located on chromosome 1 of White Leghorn chickens. J Virol 40:920–926

Tereba A, Astrin SM (1980) Chromosomal localization of *ev*-1, a frequently occurring endogenous retrovirus locus in White Leghorn chickens, by in situ hybridization. J Virol 35:888–894

Tereba A, Astrin SM (1982) Chromosomal clustering of five endogenous retrovirus loci in White Leghorn chickens. J Virol 43:737–740

Tereba A, Lai MMC (1982) Cell oncogenes are located on the large microchromosomes in chicken cells. Virology 116:654–657

Tereba A, Lai MMC, Murti KG (1979) Chromosome 1 contains the endogenous RAV-0 retrovirus sequences in chicken cells. Proc Natl Acad Sci, USA 76:6486–6490

Tereba A, Crittenden LB, Astrin SM (1981) Chromosomal localization of three endogenous retrovirus loci associated with virus production in White Leghorn chickens. J Virol 39:282–289

Traina VL, Taylor BA, Cohen JC (1981) Genetic mapping of endogenous mouse mammary tumor viruses: locus characterization, segregation, and chromosomal distribution. J Virol 40:735–744

Van Nie R, Verstraeten AA (1975) Studies of genetic transmission of mammary tumor virus by C3Hf mice. Int J Cancer 16:922–931

Verstraeten AA, Van Nie R (1978) Genetic transmission of mammary tumor virus by the DBAf mouse strain. Int J Cancer 21:473–475

Subcellular Localization of pp60^{src} in RSV-Transformed Cells

JAMES G. KRUEGER, ELLEN A. GARBER, and ALLAN R. GOLDBERG

1	Introduction	52
2	Subcellular Localization of pp60^{src}	53
2.1	In Transformed Chicken Embryo Fibroblasts	54
2.2	In SR-NRK Cells	60
2.3	In Transformed Field Vole Cells	65
3	Molecular Organization of pp60^{src} in Cellular Membranes	66
3.1	Biochemical Definition of a Membrane Protein	66
3.2	pp60^{src} Is an Integral Membrane Protein	67
3.3	The Amino-Terminus of pp60^{src} Is a Membrane-Binding Domain	68
4	Subcellular Localization of Mutant and Variant pp60^{src}s	69
4.1	Recovered Avian Sarcoma Virus-Infected CEFs	70
4.2	Temperature-Sensitive Mutant-Infected CEFs	72
4.3	RR1022 Rat Cells, Pcl Goat Cells, and RSV-RR and RSV-Pcl Viruses in CEFs	73
4.4	CU Mutant-Infected CEFs	78
5	Lipid Is Tightly Associated with pp60^{src}	78
5.1	Lipid Binding to pp60^{src} Correlates with Membrane Binding and Integration	79
5.2	Revised Model of pp60^{src} Interaction with Membranes	80
6	Localization of pp60^{src} in Adhesion Plaques	81
6.1	Introduction to Cell-Substratum Adhesions	81
6.2	Adhesion Plaques in RSV-Transformed Cells	83
7	Transport of pp60^{src} to the Membrane	88
8	Putative Substrates of pp60^{src}	91
8.1	pp60^{src} Is a Tyrosyl Kinase: Strategies for Identifying Substrates	91
8.1.1	Coimmunoprecipitation with pp60^{src}	93
8.1.2	Comparison of Phosphoproteins Resolved by Electrophoresis	93
8.1.3	Proteins of Known Function	94
8.1.4	In Vitro Systems	95
8.2	The 36-kd Transformation-Specific Phosphoprotein	96
8.3	The 46-kd and 28-kd Alkali-Resistant Putative Substrates	98
8.4	Vinculin	99
8.5	Proteins Phosphorylated in In Vitro Systems	101
9	Concluding Remarks	103
10	Appendix: General Methods Used for Subcellular Localization of Proteins	106
10.1	Cellular Fractionation	106
10.1.1	Subcellular Component Separation by Differential Centrifugation and Isopyknic Centrifugation	107
10.1.1.1	Differential Centrifugation	107
10.1.1.2	Isopyknic Centrifugation	109
10.1.2	Caveats Concerning Cell Homogenization	110
10.1.3	Analysis of Information Derived from Cellular Fractionation	112

The Rockefeller University, New York, NY 10021, USA

Current Topics in Microbiology and Immunology, Vol. 107

10.2 Microscopic Techniques for Subcellular Localization of Proteins 113
10.2.1 Immunofluorescence Microscopy 113
10.2.2 Immunoelectron Microscopy 114
10.3 Cellular Fractionation with Nonionic Detergents 115
References 116

1 Introduction

In 1911 Rous demonstrated that a retrovirus, subsequently named Rous sarcoma virus (RSV), could cause a tumor in a chicken. Nearly 60 years elapsed before it was shown that a specific virus function encoded by the *src* gene of RSV was required for the maintenance of the transformed state. Martin (1970) isolated a temperature-sensitive RSV mutant whose properties indicated that the viral transforming gene product is required for transformation, but not for viral replication. Seven years later, Martin's genetic experiment was provided with a biochemical foundation when Brugge and Erikson (1977) showed that the transformation gene *src* encoded a 60000-dalton (60-kd) protein present in RSV-transformed cells. Since 1977, many genes capable of inducing oncogenic transformation, and hence given the generic name *onc*, have been recognized, and many of the oncogenes' protein products have been identified. The molecular biology of retroviruses is reviewed extensively and in great detail by Weiss et al. (1982), and the reader is referred to this volume and numerous reviews referenced therein on specific aspects of retrovirology.

RSV transforms cells in culture both rapidly and efficiently and causes tumors in a variety of animals. The alteration of cells in vitro and the malignant changes in vivo are the consequences of a single viral gene called *src* for sarcoma (see review by Erikson 1981). Identification of a 60-kd phosphoprotein ($pp60^{src}$) as the product of the *src* gene was accomplished both by immunoprecipitation of radiolabeled polypeptides from RSV-transformed cells with an antiserum (tumor-bearing rabbit serum; TBR) raised in rabbits bearing RSV-induced tumors (Brugge and Erikson 1977; Levinson et al. 1978; Erikson et al. 1980) and by in vitro translation of 21S–24S virion RNA (Purchio et al., 1977; Beemon and Hunter, 1978). The observation that incubation of immunoprecipitates containing $pp60^{src}$ with gamma-labeled [^{32}P]ATP resulted in the phosphorylation of the heavy chain of rabbit immunoglobulin from TBR sera suggested that $pp60^{src}$ might be a protein kinase (Collett and Erikson 1978; Levinson et al. 1978). (Such a reaction is called an immune complex assay for $pp60^{src}$-associated protein kinase activity). Additional findings have demonstrated that $pp60^{src}$ possesses intrinsic protein kinase activity (see Sect. 8.1), suggesting that the pleiomorphic effects of transformation by RSV could occur via $pp60^{src}$-mediated phosphorylation of a number of cellular proteins.

It is likely that *onc* proteins transform cells by altering critical cellular control mechanisms which reside at specific subcellular sites. Thus, it may

Table 1. Altered properties of RSV-transformed cells and potential subcellular site affected by pp60 to produce observed change

Transformed Property	Potential Target
Altered transcription of genes coding for fibronectin and collagen	Nucleus
Dissolution of actin microfilament bundles, intermediate filaments, and microtubules	Cytoplasm or membrane anchor regions
Increased lactic acid production and increase in glycolytic enzymes	Cytoplasm or mitochondria
Decreased cAMP and altered adenylate cyclase activity	Plasma membrane or cytoplasm
Altered cell surface antigens and membrane constituents	Plasma membrane
Increased lectin agglutinability	Plasma membrane
Increased hexose transport	Plasma membrane
Decreased cell surface fibronectin	Extracellular matrix
Increased secretion proteases	External surface plasma membrane and pericellular space
Decreased cell adhesiveness to substratum	Plasma membrane or extracellular matrix
Increased membrane ruffling	Plasma membrane or cytoskeleton
Altered growth properties: loss of contact inhibition and increased soft agar colony formation	Nucleus, cytoplasm and/or various cellular membranes or membrane-bounded organelles

be possible to identify the site of action of a particular *onc* protein within a cell by determining its specific subcellular location. Furthermore, biochemical analysis of the interaction of *onc* proteins with specific intracellular sites may provide new insights into their mechanism of transformation. Since the greatest amount of work in this area has concentrated on pp60^{src}, the protein product of the *src* gene of Rous sarcoma virus, this review will focus on efforts to localize it and to determine its potential substrates and mechanism of action.

2 Subcellular Localization of pp60^{src}

The identification of pp60^{src} as the product of the *src* gene of RSV has made it feasible to investigate the mechanism of viral transformation. The interaction of pp60^{src} with a cell produces many changes which together comprise the transformed phenotype (Table 1) (for review, see HANAFUSA 1977; BAUER et al. 1982). Because of the pleiotropic effects of pp60^{src} on cellular processes, it has not been possible to predict a specific intracellular site or target at which pp60^{src} might act. Thus a goal of numerous laboratories during the past few years has been to determine the precise subcellular localization of pp60^{src} in several different RSV-transformed cell types. The subcellular localization of pp60^{src} will be considered separately for each cell type. For those unfamiliar with methods and problems of subcellular localization of proteins, Sect. 10 contains an introduction to and critical evaluation of these techniques.

2.1 In Transformed Chicken Embryo Fibroblasts

The subcellular localization of pp60^{src} is perhaps best studied in permissive chick embryo fibroblast (CEF) cultures, for the following reasons: (a) RSV quickly and reproducibly transforms CEF cultures, permitting study of both uninfected and transformed cells; (b) CEF cultures can be infected by a wide range of mutants temperature sensitive for transformation, as well as numerous other RSV mutants; (c) the interaction of RSV with a cell has been carefully analyzed in transformed CEFs where characteristics of the transformed phenotype are well known, as are some correlations between in vitro cellular transformation and in vivo tumorigenicity; and (d) CEF cultures can be infected with many other classes of sarcoma-type viruses and comparison of the interaction of their transforming proteins and of pp60^{src} with specific subcellular sites also can be analyzed in this cell culture system.

An early study by BRUGGE et al. (1978) using nonionic detergents to fractionate RSV-transformed CEFs into nuclear and "cytoplasmic" fractions showed that pp60^{src} was predominantly cytoplasmic. Further experiments to localize pp60^{src} within the cytoplasm were performed on cell lysates prepared by sonication and therefore provided inconclusive data on the disposition of pp60^{src} within the cytoplasm (see Sect. 10.1.1). The idea that pp60^{src} might be associated with cytosolic membranes derived from the observation that approximately four times greater amounts of pp60^{src} could be extracted from RSV-transformed CEFs by a buffered solution containing 1% Triton X-100 than by the same buffer lacking detergents (GOLDBERG et al. 1980; KRUEGER et al. 1980a). To further test whether or not pp60^{src} was associated with membranes, cellular fractionation was performed with RSV-transformed CEFs (KRUEGER et al. 1980a): cells were swollen in hypotonic buffer, Dounce homogenized, and subjected to differential centrifugation so that subcellular components were divided into four fractions: N, P_{50}, P_{200} and S_{200}. Table 2 shows the quantitative distribution of pp60^{src} in these fractions and also lists the subcellular components which were contained in each fraction. pp60^{src} was found primarily in P_{50} and P_{200} membrane-organelle fractions. The specific activity of pp60^{src} kinase was also highest in these fractions. From these data and from additional experiments in which cells were homogenized and fractionated in buffers containing varying ionic compositions (KRUEGER et al. 1980a), it was concluded that (a) it was likely that pp60^{src} was associated with cellular membranous components, (b) it was unlikely that pp60^{src} was associated with nuclei to

Table 2. Fractionation of RSV-transformed CEFs by differential centrifugation

Fraction	Cell protein (%)	pp60 (%)[a]	pp60 Specific activity[b]	NADH diaphorase (%)[c]	5′-Nucleotidase (%)[d]	Subcellular components[e]
N	13	8	0.6	8	28	Nuclei, large sheets of plasma membranes
P_{50}	36	50	1.4	76	56	Mitochondria, plasma membranes, endoplasmic reticulum, Golgi, lysosomes, Macromolecular aggregates
P_{200}	10	11	1.1	8	6	Ribosomes, microsomes derived from cellular membranes
S_{200}	41	31	0.75	8	10	Soluble cytosolic proteins and constituents

[a] Kinase activity

[b] Specific activity $= \dfrac{\%\ \text{pp60 in fraction}}{\%\ \text{cell protein in fraction}}$

[c] Marker enzyme for endoplasmic reticulum

[d] Marker enzyme for plasma membranes

[e] From enzyme marker data, electron microscopy, and data of HAY (1974)

any significant degree, and (c) it was unlikely that pp60^{src} was contained principally in the cytoplasmic ground substance as a soluble cytosolic protein.

Analysis of the immunoprecipitated radiolabeled *src* polypeptide from the S_{200} fraction showed that it had been degraded proteolytically to a lower molecular weight form (designated pp52^{src}) during homogenization or fractionation (Fig. 1). Since very low levels of pp60^{src} were observed in the soluble cytosolic fraction, KRUEGER et al. (1980a) concluded that the origin of pp52^{src} was in those pp60^{src} molecules that had been associated formerly with membranes.

To determine whether pp60^{src} was associated with a specific subcellular organelle, KRUEGER et al. (1980a) fractionated membranous subcellular components by isopyknic centrifugation on discontinuous sucrose gradients (as shown in Fig. 15 and discussed in Sect. 10.2.1). Figures 2 and 3 show the results of such an experiment. These data indicate that the majority of pp60^{src} is found in fractions 1 and 2, which are enriched for plasma membranes, as evidenced by cofractionation with 5′-nucleotidase, a marker for plasma membranes. Furthermore, the kinase specific activity is highest in those fractions. KRZYZEK et al. (1980) also investigated the subcellular location of pp60^{src} in RSV-transformed CEFs by cellular fractionation techniques and similarly concluded that pp60^{src} was associated specifically with

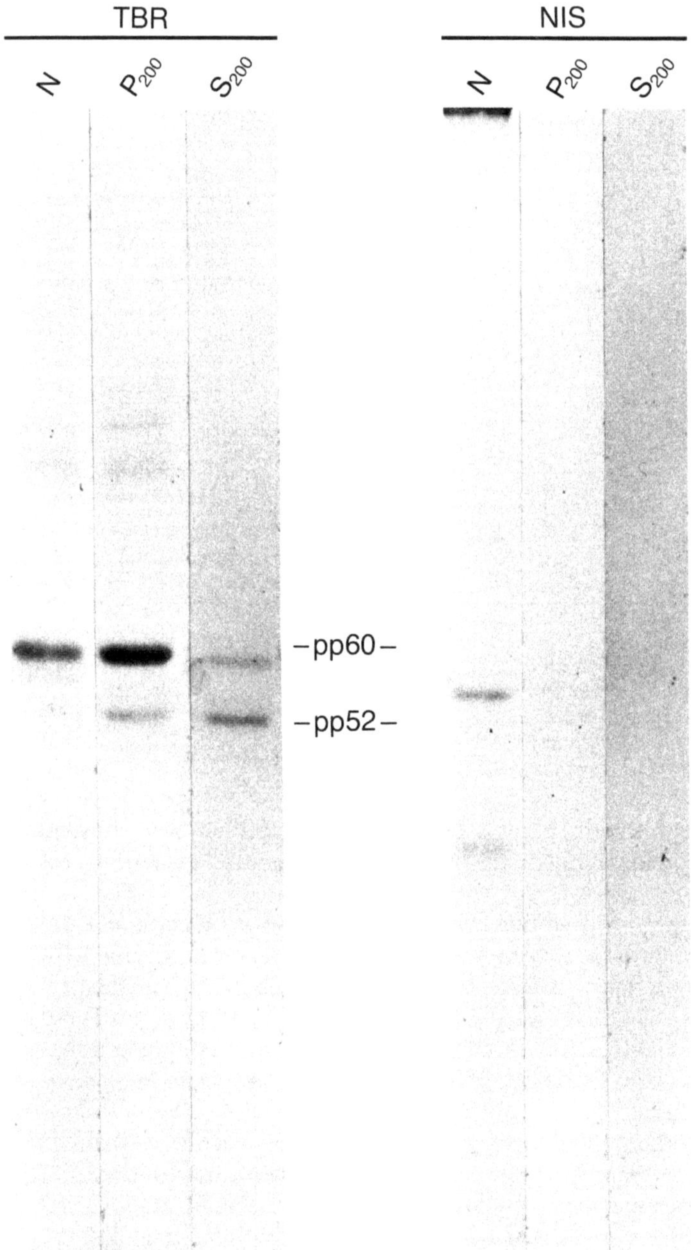

Fig. 1. Subcellular distribution of $pp60^{src}$. Nuclear (*N*), membrane (P_{200}), and soluble cytosolic (S_{200}) fractions prepared from RSV-transformed CEFs by differential centrifugation. These fractions were reacted with TBR serum to specifically immunoprecipitate $pp60^{src}$ (*left*) or with nonimmune serum (*NIS*) as a control (*right*). Cellular proteins were labeled with [^{35}S]methionine before cellular fractionation and subsequent immunoprecipitation of $pp60^{src}$

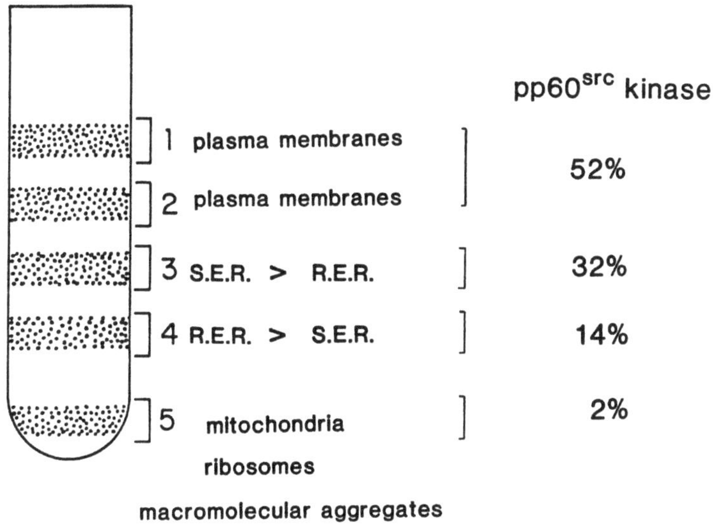

Fig. 2. Distribution of pp60^{src} in discontinuous sucrose gradients which separate subcellular membrane vesicles and organelles according to their isopyknic density (see Sect. 10.1.1 for an explanation of this fractionation scheme). pp60^{src} fractionates in parallel with plasma membranes. *S.E.R.*, smooth endoplasmic reticulum; *R.E.R*, rough endoplasmic reticulum

plasma membranes. The data of KRUEGER et al. (1980a) and KRZYZEK et al. (1980) cannot conclusively exclude association of pp60^{src} with lysosomes or Golgi membranes that often copurify with plasma membrane fractions obtained by isopyknic centrifugation; however, it is unlikely that such is the case because both Golgi membranes and lysosomes are present in low levels in cultured embryonic fibroblasts.

Indirect immunofluorescence microscopy gave visual evidence that pp60^{src} is associated with plasma membranes (KRUEGER et al. 1980b), supporting the cell fractionation data. TBR serum was freed of antibodies to viral structural proteins by affinity chromatography on agarose to which viral structural proteins had been coupled covalently. The resulting serum, and IgG purified from it, were judged to be monospecific for pp60^{src} by immunoprecipitation of radiolabeled chicken cells productively infected with RSV. Figure 4 shows the pattern of immunofluorescence obtained with this antibody after incubation with RSV-transformed CEFs. Prominent fluorescence of the entire circumference of the cell and at regions of cell-to-cell contact are evident. This membrane-specific staining pattern was not observed unless cells were made permeable to antibodies by incubation in detergent-containing solutions. The differential staining after permeabilization suggests that pp60^{src} is not significantly exposed on the ectoplasmic side of the plasma membrane, but that the antigenic sites in pp60^{src} are concentrated on the cytoplasmic face of the plasma membrane. These data do not exclude the possibility that a small amount of pp60^{src} is exposed on the external surface of the cell, since there may not be strong antigenic sites in this region of the molecule.

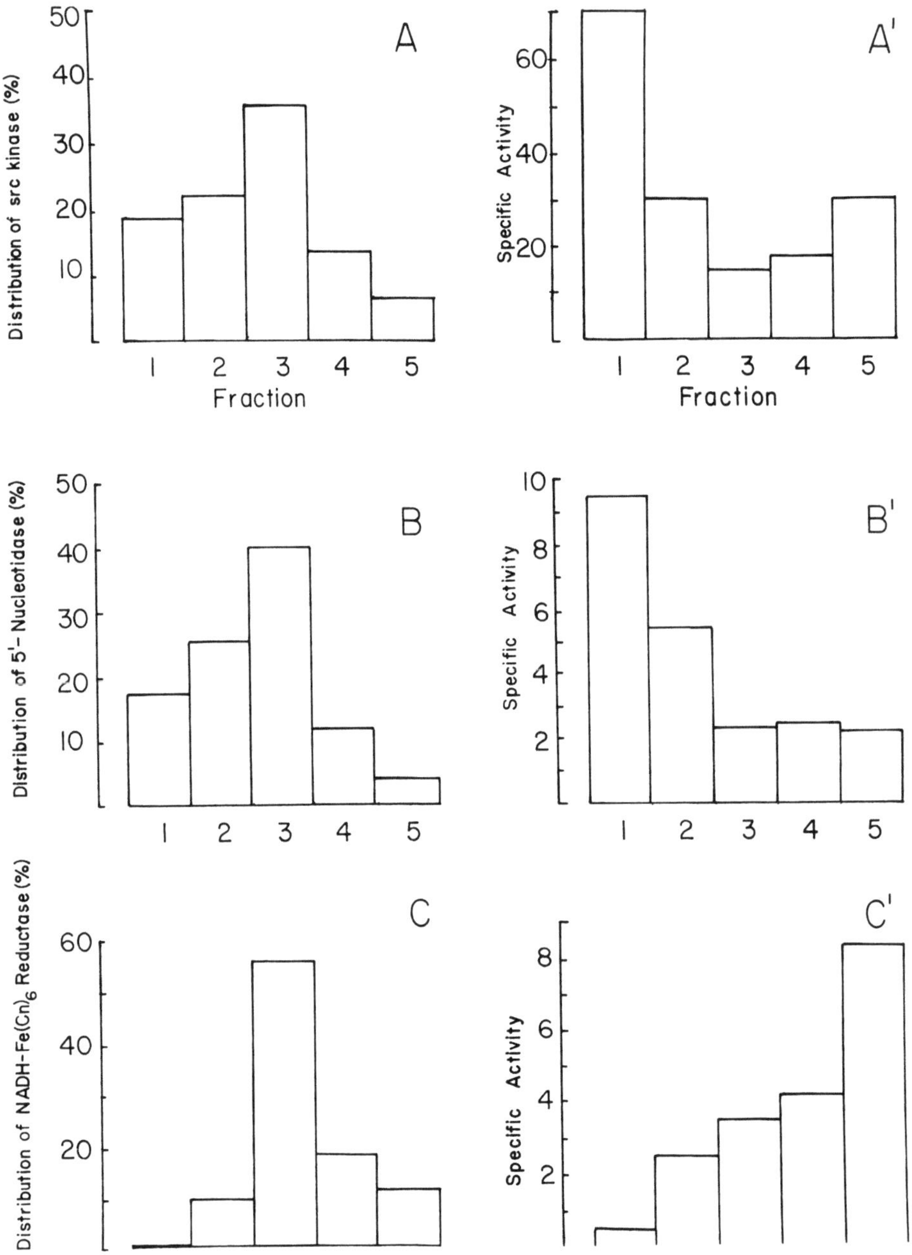

Fig. 3. Distribution of pp60^{src} (*A*), 5′-nucleotidase (*B*), and NADH diaphorase (*C*) in P_{50} fractions of RSV-transformed CEFs which were centrifuged in isopyknic sucrose gradients. 5′-Nucleotidase is a marker for plasma membranes; NADH diaphorase is a marker for endoplasmic reticulum. Panels (*A′*), (*B′*), and (*C′*) show the specific activities of each enzymatic activity

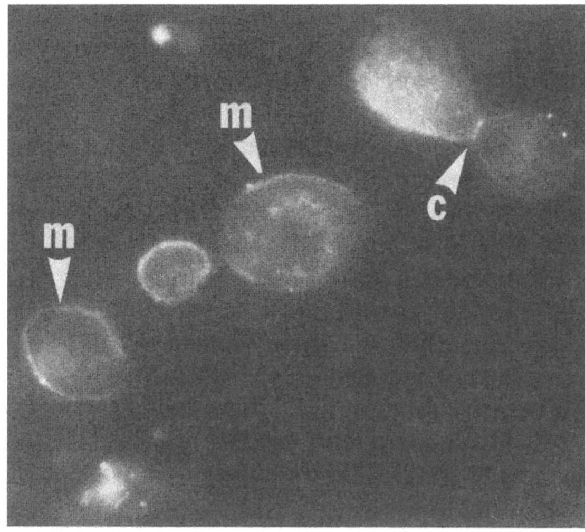

Fig. 4. Localization of pp60^{src} in SR-RSV-A-transformed CEFs by indirect immunofluorescence microscopy. Cells were stained for pp60^{src} using TBR serum from which all antibodies directed against viral structural proteins had been removed by affinity chromatography. pp60^{src} is present in plasma membranes (*m*) and at regions of cell-to-cell contact (*c*)

An earlier study using immunofluorescence microscopy to localize pp60^{src} in RSV-transformed CEFs showed that pp60^{src} was present either at regions of cell-to-cell contact or in the cytoplasm of transformed cells. The cytoplasmic staining was visualized either as a cap of perinuclear fluorescence or as a diffuse cytoplasmic fluorescence (ROHRSCHNEIDER 1979). The cytoplasmic pattern of fluorescence might be the result of association of pp60^{src} with either intracellular organelles or soluble cytosolic components. It is difficult to distinguish between these possibilities in this instance, because cells were fixed and made permeable by treatment with acetone, thereby altering the integrity of intracytoplasmic structures (WILLINGHAM et al. 1979). The differences in the immunofluorescence patterns observed by ROHRSCHNEIDER (1979) and KRUEGER et al. (1980b) in RSV-transformed CEFs may be due to differences in antibody preparation, fixation, and permeabilization procedures or in the strain of virus used to transform the cultures. Nonetheless, both immunofluorescence studies support the non-nuclear subcellular location of pp60^{src} and suggest that pp60^{src} can be found in cellular plasma membranes at regions of cell-to-cell contact. Joint consideration of immunofluorescence data and cellular fractionation data support the conclusion that pp60^{src} is associated with plasma membranes in RSV-transformed CEFs.

Several different laboratories have posed the question of whether any portion of pp60^{src} is expressed on the surface of transformed cells. ROHRSCHNEIDER (1979) could not detect surface fluorescence on either viable chicken or rat cells transformed by the Schmidt-Ruppin subgroup D strain of RSV (SR-RSV-D). However, using pp60^{src}-specific serum, BARNEKOW et al. (1981, 1982) observed surface fluorescence on fixed, nonpermeabilized SR-RSV-transformed CEFs. The cell surface antigen was identified as pp60^{src}, not an unrelated transformation-specific antigen, by demonstrating

$pp60^{src}$ kinase activity and ^{32}P-labeled $pp60^{src}$ on the surface of transformed cells by Triton extraction of cells previously exposed to antibody. These experiments do not rule out the possibility that only a small portion of the $pp60^{src}$ molecule is exposed on the cell surface. These workers also detected release of $pp60^{src}$ kinase activity from transformed cells into the incubation medium. This release may be due to budding of virions from the plasma membrane or shedding of membrane fragments containing $pp60^{src}$. BUNTE et al. (1981) and OWADA et al. (1981) detected $pp60^{src}$ in the membranes of virions, but found that $pp60^{src}$ in intact virus particles was inaccessible to antibodies, surface iodination, or mild proteolysis, thereby suggesting that the $pp60^{src}$ included in the virus during the budding process was not exposed on the cell surface.

Subsequent studies by ROHRSCHNEIDER (1980) and ROHRSCHNEIDER et al. (1982) using indirect immunofluorescence techniques showed that $pp60^{src}$ was concentrated in adhesion plaques in RSV-transformed CEFs. This localization is not inconsistent with a primary plasma membrane location of $pp60^{src}$, because adhesion plaques form in or near plasma membranes. Localization of $pp60^{src}$ in adhesion plaques in these and other cell types is discussed further in Sect. 6.

BURR et al. (1980, 1981) found that most of the protein kinase activity and ^{32}P-labeled $pp60^{src}$ were associated with the cytoskeletal framework after extraction of RSV-transformed CEFs with buffers containing nonionic detergents. These authors report their belief that this localization reflects the true in vivo state of $pp60^{src}$ and was not the result of artifactual binding. The limitations of localization using detergent extraction are discussed in Sect. 10.3. The finding that $pp60^{src}$ is associated with the cytoskeletal fraction is not inconsistent with $pp60^{src}$ being an integral membrane protein associated (directly or indirectly) with elements of the cytoskeleton (see, for example, BEN-ZE'EV et al. 1979).

2.2 In SR-NRK Cells

The localization of $pp60^{src}$ in normal rat kidney cells transformed by the Schmidt-Ruppin subgroup D strain of RSV, SR-NRK cells, has been investigated by several different laboratories using the techniques of cell fractionation, indirect immunofluorescence, and immunoelectron microscopy. These cells have two distinct advantages over RSV-transformed CEFs. First, the cells are a stable tissue culture line which can be propagated easily in tissue culture by serial passage, and thus normal cells do not need to be transformed for every experiment. Second, these cells constitute a more homogeneous population of cells than do RSV-transformed CEFs. However, there also are some distinct problems in studying SR-NRK cells. Because this cell line has been maintained in tissue culture for many years in different laboratories, these cells may have acquired new properties which may be unrelated to the transforming influence of $pp60^{src}$. In fact, one cannot be sure that cell transformation is wholly attributable to $pp60^{src}$, since many cells undergo spontaneous transformation on continued passage in vitro. In consideration of the divergence of growth properties on continued passage, it may not be fair to compare SR-NRK cells with the population of NRK cells from which they were derived. The SR-NRK cells used for pp60^{src} localization have different passage histories in different laboratories and may be substantially different from each other in their properties.

ROHRSCHNEIDER (1979) first reported that pp60^{src}, visualized by indirect immunofluorescence, was localized at or near plasma membranes at regions of cell-to-cell contact in dense cultures of SR-NRK cells. However, in both sparse and dense cultures, the predominant pattern of fluorescence was cytoplasmic. In this context "cytoplasmic" refers to a non-nuclear location of pp60^{src}, but does not distinguish between pp60^{src} association with intracytoplasmic organelles and membranes and location of pp60^{src} within the soluble phase of the cytosol. Such a distinction cannot justifiably be made when cells have been fixed and made permeable with acetone in preparation for immunofluorescence microscopy (see Sect. 10.2.1). In a subsequent study, WILLINGHAM et al. (1979) showed by immunofluorescence microscopy that pp60^{src} was localized at regions of cell-to-cell contact, at the free edge of the cell (plasma membrane), and in a perinuclear cytoplasmic location (Fig. 5). For more precise localization of pp60^{src}, immunoelectron microscopy was performed with the EGS procedure (see Sect. 10.2.2). Those experiments indicate that pp60^{src} was positioned at the inner or cytoplasmic surface of the plasma membrane with its greatest concentration at specialized plasma membrane gap junctions (Fig. 6). While the concentration of ferritin cores, reflective of the pp60^{src} concentration (see Fig. 16), was quite low in the cytoplasm seen in individual thin sections, the total number of intracellular cores was quite large when the total cytoplasmic volume was calculated. WILLINGHAM et al. (1979) calculated that up to 60% of pp60^{src} could be found in the cytosol of SR-NRK cells. However, they suggest that the concentration of pp60^{src} at a specific subcellular location may be the most important factor in determining its major site of action. The correspondence of the immunofluorescence and immunoelectron microscopic localizations of pp60^{src} is not perfect. The perinuclear cytoplasmic fluorescence observed in these cells suggests that pp60^{src} may be associated with a discrete intracytoplasmic structure located in a perinuclear position. A soluble cytosolic protein would not be expected to show a discrete cytoplasmic location. Immunoelectron microscopy did not show any specific association of pp60^{src} with perinuclear structures, but suggested that it was largely cytoplasmic. The immunofluorescence and immunoelectron microscopy studies are in better agreement on the plasma membrane location of pp60^{src}. Furthermore, the studies of both ROHRSCHNEIDER (1979) and WILLINGHAM et al. (1979) suggest that antigenic sites of pp60^{src} are located on the inner or cytoplasmic surface of the plasma membrane and that no such sites are exposed on the external cell surface.

COURTNEIDGE et al. (1980) localized pp60^{src} in SR-NRK cells by subcellular fractionation. Their techniques were modified from HAY (1974) and the comments on these techniques in Sects. 10.1.1–10.1.3 are relevant to

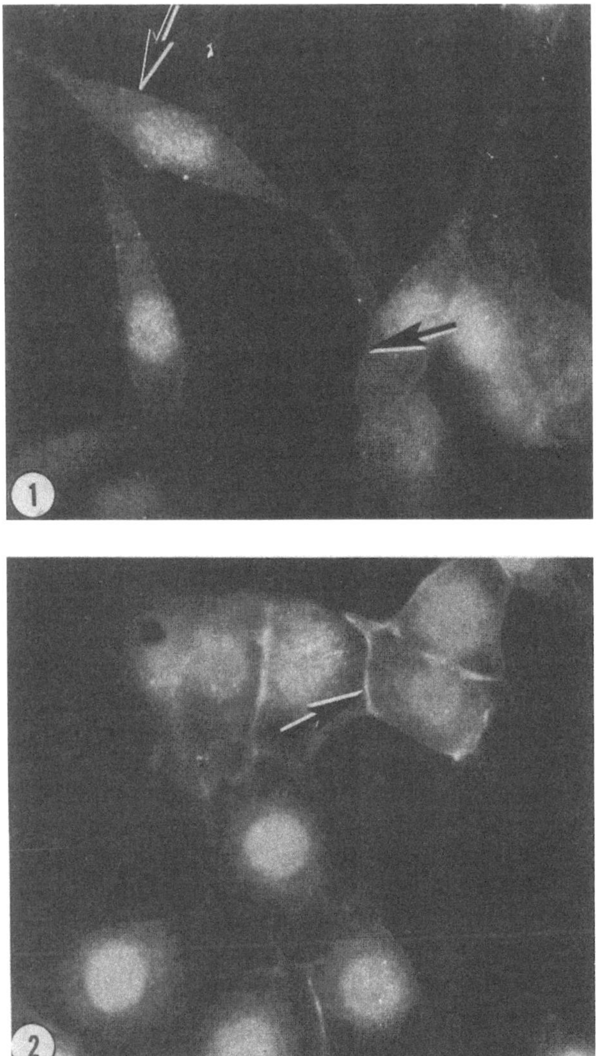

Fig. 5. Localization of pp60^{src} in SR-NRK cells by indirect immunofluorescence microscopy. In *1*, pp60^{src} can be seen in the free cell edge; in *2*, it is clearly seen in plasma membranes and at regions of cell-to-cell contact. Plasma membrane fluorescence is indicated by arrows. (Micrographs courtesy of M. Willingham)

interpretation of their data. Dounce-homogenized cells were fractionated by differential centrifugation and the presence of pp60^{src} in each fraction was assayed. Results from these experiments are shown in Table 3. The level of pp60^{src} and pp60^{src} kinase activity in each of the fractions is not precisely the same, but a similar pattern of distribution is apparent. Most of the pp60^{src} is contained in P_{100} membrane-organelle fractions, and the kinase specific activity is highest in these fractions. It is also clear that

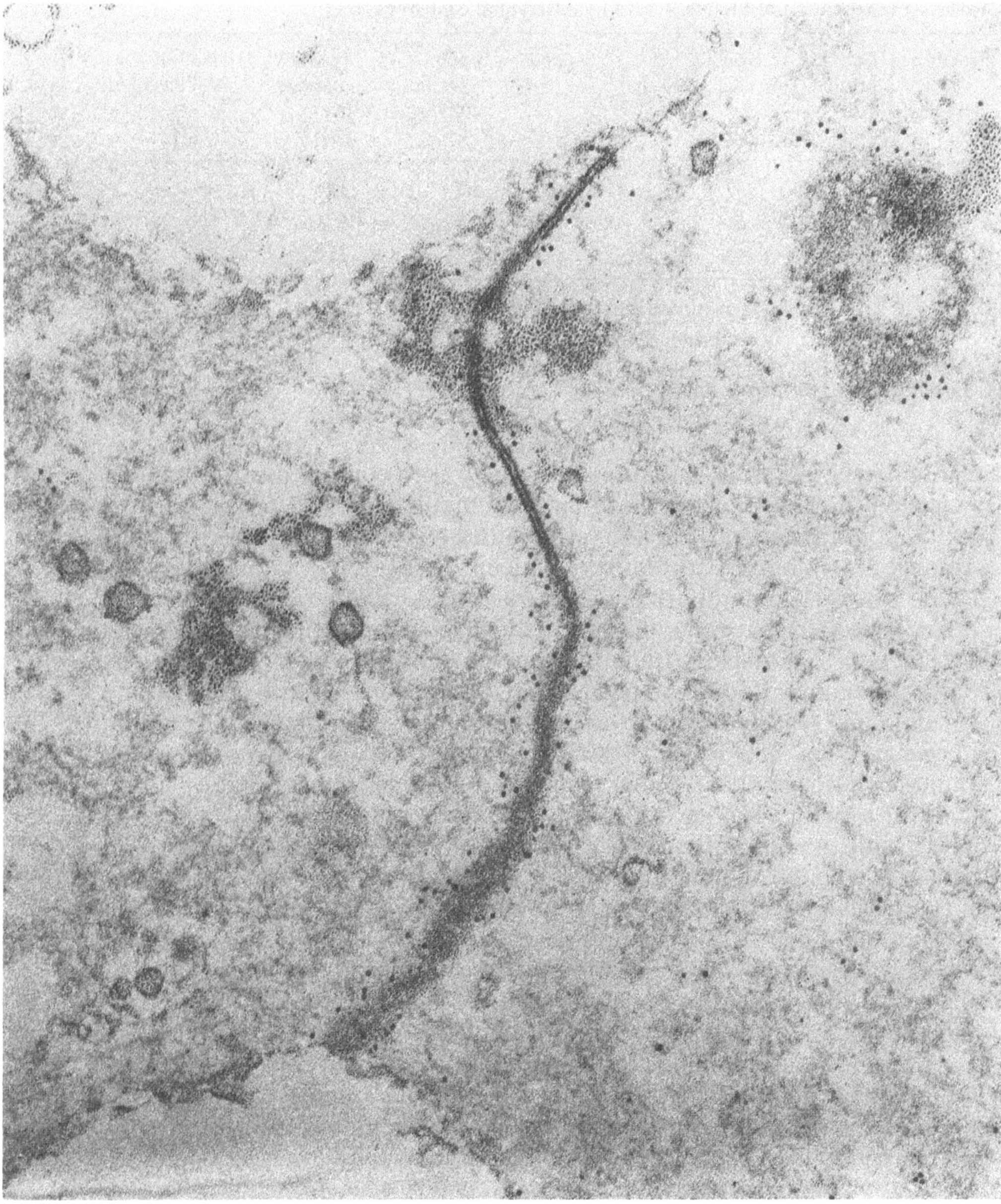

Fig. 6. Localization of pp60^{src} in SR-NRK cells by immunoelectron microscopy. Small cores of ferritin mark the position of anti-pp60^{src} antibodies in the plasma membrane between the two cells in this micrograph (see Sect. 10.2.2 for explanation of this technique). pp60^{src} is more highly concentrated in membrane junctions than nonjunctional regions of the plasma membrane. (Micrograph courtesy of M. Willingham)

Table 3. Fractionation of SR-NRK cells by differential centrifugation

Fraction	Cell protein (%)	pp60 Kinase (%)	^{35}S-pp60 (%)	pp60 Specific activity[a]	NADH diaphorase (%)[b]	Na^+/K^+ ATPase (%)[c]
N	32	37	15	1.2	32	38
P_{100}	23	51	83	2.2	45	50
S_{100}	45	12	2	0.3	23	12

[a] Specific activity $= \dfrac{\text{\% pp60 kinase in fraction}}{\text{\% cell protein in fraction}}$
[b] Enzyme marker for endoplasmic reticulum
[c] Enzyme marker for plasma membranes

quantitatively insignificant amounts of pp60^{src} are contained in the soluble cytosolic (S_{100}) fraction. Because large amounts of pp60^{src} kinase activity are found in the nuclear fraction, association of pp60^{src} with nuclei or with perinuclear membranes cannot be excluded. While a lesser amount of intracellular pp60^{src} is found in the nuclear fraction, earlier immunofluorescence studies suggested that some pp60^{src} was localized in a perinuclear cytoplasmic location. The nuclear fraction prepared from SR-NRK cells contains a large percentage of total cell protein and is contaminated with endoplasmic reticulum and plasma membranes, based upon the fractionation of specific enzyme markers. This nuclear fraction would presumably also contain contaminating lysosomal and Golgi membranes and, since these organelles are found in a perinuclear location in undisrupted cells, some association of pp60^{src} with these membranes is a possibility. This finding could reconcile these data with the immunofluorescence data showing pp60^{src} in a perinuclear cytoplasmic location. These cell fractionation data provide convincing evidence that pp60^{src} is not contained in the soluble cytosol to a significant extent, contradicting the finding of WILLINGHAM et al. (1979) that up to 60% of pp60^{src} could be located in the cytosol. The reasons for the discrepancy between these two techniques is not immediately obvious, but there are many reasons why immunoelectron microscopy could provide misleading information. Quantitation of the amount of pp60^{src} contained within the soluble cytosol is best accomplished with cell fractionation followed by specific immunoprecipitation of pp60^{src}. Immunoprecipitation allows both quantitation and positive identification of pp60^{src}, whereas immunoelectron microscopy could permit simultaneous localization of unrelated cross-reacting antigens or degradation products of pp60^{src}.

COURTNEIDGE et al. (1980) established that the majority of intracellular pp60^{src} was contained in a P_{100} membrane-organelle fraction and investigated these subcellular components further by isopyknic separation. pp60^{src} was found to fractionate in parallel with plasma membranes (as determined by 5′-nucleotidase activity). This fractionation pattern was dissimilar to that of endoplasmic reticulum (as determined by NADH diaphorase activity). These findings, in association with the immunoelectron and immunofluorescence microscopy studies cited earlier, strongly suggest that $pp60^{src}$ is associated with plasma membranes in SR-NRK cells. These data, however, provide no quantitative determination of the amount of intracellular $pp60^{src}$ that is found in plasma membranes, because $pp60^{src}$ could also be associated with other intracellular membranes which have a density similar to plasma membranes (e.g., Golgi and lysosomal membranes). The claim of COURTNEIDGE et al. (1980) that most, if not all, of $pp60^{src}$ is located in the plasma membrane is probably an oversimplification of their data, especially when one considers the immunofluorescence data contrary to this point of view. Further experiments are necessary to fully resolve this issue.

In subsequent immunofluorescence studies, ROHRSCHNEIDER and coworkers (ROHRSCHNEIDER 1980; SHRIVER and ROHRSCHNEIDER 1981a, b; ROHRSCHNEIDER et al. 1982) also demonstrated the presence of $pp60^{src}$ in adhesion plaques in SR-NRK cells. This location is not inconsistent with a primary location of $pp60^{src}$ in plasma membranes and the significance of this finding is considered in Sect. 6.

SCHAFFHAUSEN et al. (1982) reported that $pp60^{src}$ is associated with the cytoskeletal framework of SR-NRK cells remaining after extraction with nonionic detergents. As discussed in Sect. 2.1, the localization of a protein to the cytoskeletal fraction of cells can be reconciled with its being an integral membrane protein.

2.3 In Transformed Field Vole Cells

Morphological revertants of RSV-transformed field vole cells have been characterized. These cells are of particular interest because the mutation responsible for the morphological reversion to an untransformed phenotype is thought to reside in cellular targets, since virus rescued from revertant cells shows normal transforming potential in CEF cultures (LAU et al. 1979, 1981). KRZYZEK et al. (1980) reasoned that these cellular mutations might interfere with the ability of $pp60^{src}$ to interact with specific intracellular targets. For those reasons, these workers investigated the location of $pp60^{src}$ in parental and revertant RSV-transformed cells by cellular fractionation. Their results showed that over 80% of $pp60^{src}$ was associated with particulate subcellular components in both transformed and revertant cells. Isopyknic fractionation of P_{50} subcellular membranes and organelles suggested that $pp60^{src}$ was associated with plasma membranes to the same extent in transformed and revertant cells. These investigators suggested, therefore, that morphological reversion might be due to loss or alteration of a plasma membrane target protein with which $pp60^{src}$ normally interacts. Cellular mutations of this type are likely to be key tools in elucidating the specific function of $pp60^{src}$ within cellular membranes. The study of such cellular mutations is best accomplished in transformed and revertant mammalian cells, since uninfected avian cells do not form stable tissue culture lines. It is worth noting that field vole plasma membranes and lysosomes cofractionated in isopyknic gradients, as determined by the distribution of 5′-nucleotidase and acid phosphatase respectively; pp60^{src} could therefore be associated to some extent with lysosomal membranes in these cells.

3 Molecular Organization of pp60^{src} in Cellular Membranes

Investigation of the biochemical details of the interaction of pp60^{src} with cellular membranes is considered in this section and in Sect. 5. This section deals with data derived from experiments with wild-type pp60^{src}. Section 4 deals principally with cellular fractionation data of mutants and variants of wild-type pp60^{src}, and Sect. 5 deals with an additional mechanism for pp60^{src} membrane association based upon experiments with wild-type, variant, and mutant pp60^{src}s. These mutants are introduced in the next section. This section begins with a brief introduction to biochemical definitions of membrane and cytosolic proteins.

3.1 Biochemical Definition of a Membrane Protein

A protein may be associated with a cellular membrane as either a peripheral or an integral protein. Peripheral proteins usually associate with membranes through relatively weak ionic interactions, whereas integral proteins bind strongly to membranes through hydrophobic interactions. As discussed by Singer (1974), it is not always easy to decide whether a particular protein is integral or peripheral, since integral membrane proteins have heterogenous properties. Nonetheless, it is often important to be able to make this distinction, because integral membrane proteins are frequently synthesized by different pathways from peripheral proteins (Blobel 1980), and ultimate biological and biochemical properties are influenced by the hydrophobic versus hydrophilic nature of proteins. The delineation between the integral or peripheral nature of a protein is often operational and is decided to a large extent by its behavior in cell fractionation. The criteria for making this distinction are shown in Table 4.

It is more difficult to define those proteins which bind to membranes in hypotonic buffers but which are largely released from membranes in isotonic buffers. These proteins are usually classified as soluble cytosolic proteins, since weak ionic interactions can cause redistribution of cellular proteins during cellular fractionation (Emmelot and Bos 1966). Cytosolic proteins such as lactate hydrogenase do not associate with cellular membranes even in low ionic strength buffers (Feldman et al. 1983). Nonetheless, the definition of a protein's solubility by cellular fractionation is somewhat artificial since the ionic constituents of cellular fractionation buffers differ from those of the cytosol and the cytosol is diluted 10- to 100-fold during cell fractionation procedures. A protein which is classified as a soluble cyto-

Table 4. Biochemical properties of integral and peripheral membrane proteins

Property	Peripheral protein	Integral protein
Fractionates with cellular membranes in differential centrifugation of cell homogenates	Usually	+
Removed from membranes by iso-osmotic or hyperosmotic salt concentrations in buffers	+	−
Removed from membranes by divalent cation chelators (e.g., EDTA, EGTA)	+ or −	−
Removed from membranes by protein-denaturing agents (e.g., urea, pH extremes)	+	−
Requires detergents for release from membranes or to solubilize into aqueous buffers	−	+
Purified protein contains bound phospholipid or detergent	−	+

solic protein by cellular fractionation might interact with cellular membranes as a peripheral protein in undisrupted cells. Immunofluorescence and immunoelectron microscopy also can be useful tools for the determination of a protein's interaction with membranes in undisrupted cells.

3.2 pp60^{src} Is an Integral Membrane Protein

On the basis of the criteria listed in Table 4, pp60^{src} should be classified as an integral membrane protein. KRUEGER et al. (1980a) showed that pp60^{src} fractionated with cellular membranes and was not removed from these membranes by incubation with buffers containing hypertonic NaCl or $MgCl_2$. It was also not removed from membranes by the divalent cation chelators EDTA or EGTA. In addition, incubation of membranes with buffers containing 4 *M* urea did not release pp60^{src} from membranes (KRUEGER, unpublished work), whereas nonionic detergents released pp60^{src} from membranes and were required to solubilize pp60^{src} in aqueous buffers (KRUEGER et al. 1980a). LEVINSON et al. (1981) also have demonstrated that pp60^{src} attached to cellular membranes cannot be extracted with hypertonic KCl buffers or buffers containing divalent cation chelators, whereas it is solubilized by buffers containing nonionic detergents. Furthermore, pp60^{src} contains bound lipid (SEFTON et al. 1982b) which is attached to the putative membrane-binding domain of pp60^{src} (GARBER et al. 1983b). These experiments strongly support the suggestion that pp60^{src} is an integral plasma membrane protein in avian and mammalian cells infected with RSV.

3.3 The Amino-Terminus of pp60^{src} Is a Membrane-Binding Domain

A 52-kd subfragment of pp60^{src} is found in soluble cytosolic fractions prepared from RSV-transformed CEF cultures (Krueger et al. 1980a). This fragment, termed pp52^{src}, is antigenically related to pp60^{src}, contains *src*-associated kinase activity, and is solubilized in aqueous buffers lacking detergents. Krueger et al. (1980a) postulated that pp52^{src} is generated from pp60^{src} by an artifactual proteolytic cleavage during cellular fractionation and that the 8 kd removed from pp60^{src} by this cleavage contains a hydrophobic membrane-binding region. Subsequent analysis of pp52^{src} by limited *Staphylococcus aureus* V8 proteolysis showed that the 8 kd is derived from the amino-terminus of pp60^{src} (Krueger et al. 1980b). It had been demonstrated previously that hydrophilic domains of proteins can be released from membranes following proteolytic cleavage (Singer 1974). A model for interaction of pp60^{src} with cellular membranes was constructed from these data (Krueger et al. 1980b) and is reproduced in Fig. 7a. This model is consistent with the organization of most integral membrane proteins into hydrophobic and hydrophilic regions or domains (Singer 1974). The observation that nonionic detergents solubilize pp60^{src} from membranes indicates that the amino-terminus of pp60^{src} is hydrophobic and interacts directly with membrane lipids.

Additional evidence for the existence of an amino-terminal membrane-binding domain of pp60^{src} is provided by the experiments of Levinson et al. (1981). Membrane vesicles containing pp60^{src} were treated with trypsin, causing limited hydrolysis of attached proteins. A soluble fragment released by this protease treatment was recovered and shown by immunoprecipitation and subsequent analysis by SDS-PAGE to be a 47-kd carboxy-terminal polypeptide derived from pp60^{src}. Additional experiments showed that the kinase-enzymatic domain is localized to a 30-kd tryptic peptide from the carboxy-terminal portion of the *src* protein. The model of pp60^{src} which was generated from these data is shown in Fig. 7b.

The models of pp60^{src} shown in Fig. 7a, b differ slightly in the size of the proposed amino-terminal membrane-binding domain: 7a shows an 8-kd domain, 7b a 13-kd domain. Recent experiments argue for the smaller size of this membrane-binding region (Krueger et al. 1982): Two isolates of recovered avian sarcoma viruses (see Sect. 4.1) encode size-variant pp60^{src}s which have alterations within the amino-terminal 8-kd region. These two isolates have greatly reduced membrane-binding ability and behave as soluble, cytosolic proteins in cellfractionation studies (see Sect. 4.1 for details). Even though the models shown in Fig. 7a, b differ slightly in the size of the postulated membrane-binding domain, both suggest that pp60^{src} binds strongly to membranes through a hydrophobic amino acid sequence. However, analysis of the nucleotide sequence of a cDNA clone of v-*src* did not show long, uninterrupted stretches of hydrophobic residues within the amino-terminal 13 kd of pp60^{src} (Takeya and Hanafusa 1982; Takeya et al. 1982). It is possible that the secondary structure of pp60^{src} may generate a large hydrophobic region, but there is no evidence to support this

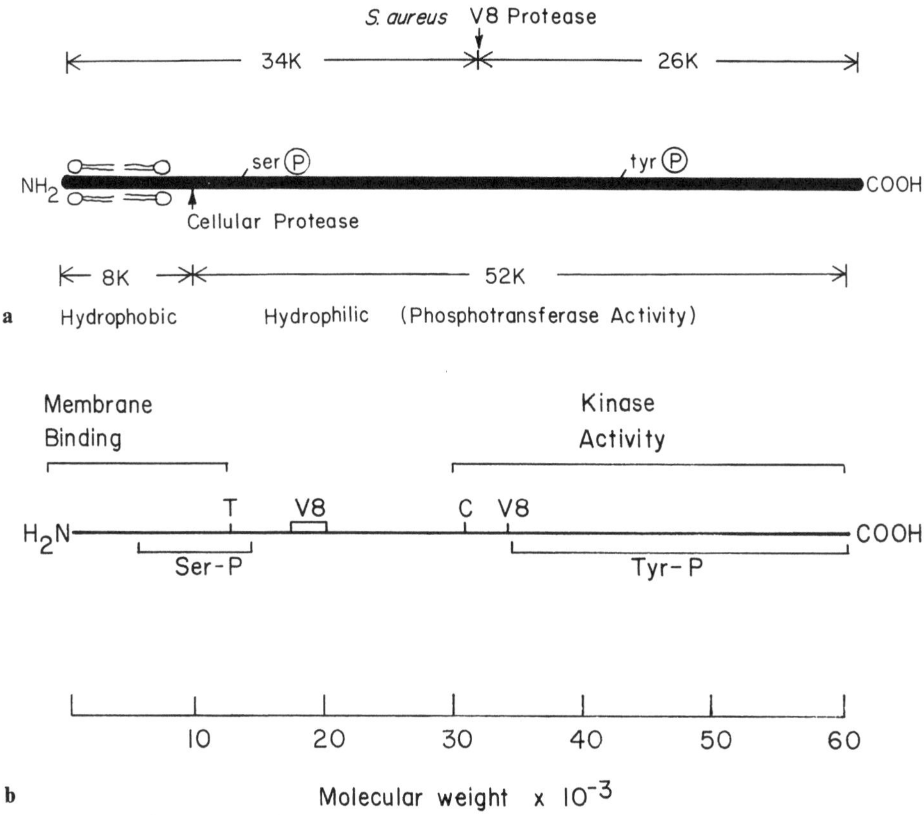

Fig. 7. a Model of pp60^{src} showing two functional domains. The amino-terminal 8-kd fragment contains one or more hydrophobic regions which allow pp60^{src} to interact directly with membrane lipids. Given this orientation, pp60^{src} could be synthesized on free or membrane-bound polysomes. During cellular fractionation, a protease present in the lysate can cleave pp60^{src} at the indicated point to release a water-soluble fragment (pp52^{src}) which retains the kinase activity. Several other features of pp60^{src} are indicated. **b** Model showing functional domains of pp60^{src}. The diagram represents approximate locations within pp60^{src} of phosphorylated serine (*Ser-P*), phosphorylated tyrosine (*Tyr-P*), a preferred site for cleavage by trypsin (*T*), a preferred site for cleavage by chymotrypsin (*C*), preferred sites for cleavage by the V8 protease (*V8*), the molecular domain which binds pp60^{src} to the plasma membrane, and the molecular domain which carries the protein kinase activity

hypothesis. Additional mechanisms to account for the hydrophobic interaction of pp60^{src} with cellular membranes are discussed in Sect. 5.

4 Subcellular Localization of Mutant and Variant pp60^{src}s

Subcellular localization of wild-type pp60^{src} in avian and mammalian cells suggested that pp60^{src} was largely, if not exclusively, located in cellular plasma membranes. This finding led to the hypothesis that some alteration

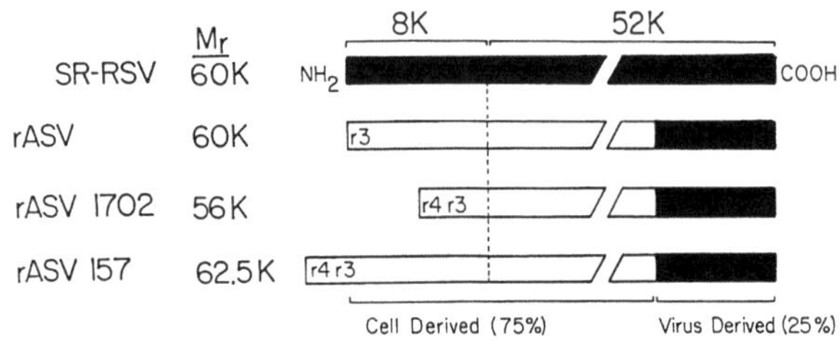

Fig. 8. Structure of rASV pp60^{src}s. Schematic drawings (not to scale) show the molecular structure of the pp60^{src}s encoded by SR-RSV and td109-derived rASVs. *r3, r4;* tryptic peptides present in the amino-terminal 8-kd region identified by KARESS and HANAFUSA (1981)

in the normal physiological function of the plasma membrane might be responsible for cellular transformation. A prediction of this hypothesis is that altered interaction of pp60^{src} with plasma membranes might diminish the ability of pp60^{src} to produce cellular transformation. Accordingly, numerous mutant and variant pp60^{src}s have been screened for altered intracellular localizations. Identification of mutants showing altered subcellular localizations should provide unique tools for studying the physiological function of pp60^{src} in cellular plasma membranes.

4.1 Recovered Avian Sarcoma Virus-Infected CEFs

Recovered avian sarcoma viruses (rASVs) were generated by in vivo recombination between the c-*src* gene and transformation-defective RSV strains with partial deletions of the v-*src* gene (HANAFUSA et al. 1977). Numerous rASVs have been generated from *td*RSV mutants with deletions of different regions of the *src* gene. In RSV *td*109-derived rASVs, the amino-terminal three-quarters of pp60^{src} is specified by c-*src* encoded sequences and the carboxy-terminal quarter is specified by v-*src* sequences (KARESS and HANAFUSA 1981). Figure 8 shows the structure of 60-kd pp60^{src}s encoded by rASVs of this class and of size-variant 56-kd and 62.5-kd pp60^{src}s encoded by rASV 1702 and rASV 157 respectively. KRUEGER et al. (1982) examined the subcellular localization of size-variant pp60^{src}s. Because the addition or deletion of amino acids responsible for the size variation mapped exclusively within 8 kd of the amino-terminal of wild-type pp60^{src} (KARESS and HANAFUSA 1981), in the region implicated as a membrane-binding domain (see Sect. 3.2), KRUEGER et al. (1982) surmised that the size-variant pp60^{src}s might show altered membrane-binding characteristics. Since rASVs are replication competent in CEF cultures, the localization of 60-kd and size-variant pp60^{src}s could be studied in the same cell culture systems, thus minimizing artifacts inherent in comparing pp60^{src} localization in different cell types. Twelve different *td*109-derived rASV isolates were used to transform CEF cultures (KRUEGER et al. 1982). Ten of the rASV stocks encoded 60-kd pp60^{src} proteins, while two (rASV 1702 and rASV 157) encoded size-variant *src* proteins. All 60-kd *src* polypeptides were found predominantly within the P_{100} (membrane-organelle) fractions obtained by differential centrifugation of cell lysates. Plasma membranes prepared by isopyknic centrifugation from cells infected by selected rASV isolates contained levels of pp60^{src} kinase comparable to that in wild-type RSV-infected cells. In contrast, the 56-kd protein encoded by rASV 1702 and the 62.5-kd protein encoded by rASV 157 were contained principally in S_{100} fractions of cellular homogenates prepared by differential centrifugation and thus appeared to fractionate as soluble cytosolic proteins. Consistent with this finding, plasma membranes prepared from cells transformed by these size-variant rASVs contained less than 10% of the amount of pp60^{src} found in membranes purified from cells transformed by RSV or control rASVs. Thus, changes in the postulated membrane-binding domain of pp60^{src} dramatically affected the ability of pp60^{src} to bind to plasma membranes. Since the association of rASV 1702 and rASV 157 pp60^{src}s with cellular membranes was influenced by the ionic composition of the fractionation buffer, it was not clear whether these pp60^{src}s might interact with plasma membranes as peripheral membrane proteins in undisrupted cells. Accordingly, immunofluorescence microscopy was used to localize rASV 1702 and rASV 157 pp60^{src}s in fixed cells (KRUEGER et al., to be published). These studies indicated that pp60^{src} was present in regions of cell-to-cell contact and in adhesion plaques in CEF cultures infected with both of these rASV isolates. The adhesion plaque localization of pp60^{src} in these cells is discussed further in Sect. 6. These data suggest that pp60^{src}s which fractionate as soluble cytosolic proteins may retain some ability to interact with cellular membranes as peripheral membrane proteins. It is not possible to quantitate the amount of membrane-attached versus soluble pp60^{src} in these cells by immunofluorescence techniques, because the soluble form of pp60^{src} may be lost from cells during processing for microscopy.

As discussed in Sect. 3.1, the membrane interaction of a protein which fractionates as a soluble protein in differential centrifugation is biochemically different from that of an integral membrane protein. Several properties of transformed cells that are thought to be related to membrane phenomena were investigated in CEF cultures transformed by normal and size-variant pp60^{src}-encoding rASVs in order to determine the effect of the lack of integration of the size-variant pp60^{src}s into plasma membranes. The ability of pp60^{src}s encoded by rASV 1702 and rASV 157 to effect cytoskeleton disorganization, to increase deoxyglucose transport across plasma membranes, to promote cell growth in soft agar, and to increase phosphorylation of a 36-kd membrane-associated cellular protein appeared to be unaltered with respect to control rASVs encoding pp60^{src}s which were integrated into membranes. However, rASV 1702- and rASV 157-transformed cells were less rounded and more elongated than control rASV-infected cells. This morphological difference may be attributable to altered cell-substratum adhesions or to retained cell surface fibronectin (see Sect. 6). The ability of rASV 1702 and rASV 157 pp60^{src}s to produce in vivo neoplastic cellular transformation, a stringent test of transforming ability, was reduced significantly compared with RSV or rASVs encoding pp60^{src}s integrated into cellular membranes. Tumors produced by rASV 1702 are benign, noninvasive neoplastic growths which regress, whereas tumors produced by RSV are malignant, invasive sarcomas which do not regress and are lethal to their hosts in a short time. It is possible that the altered membrane association of the size-variant pp60^{src}s restricts their ability to interact with some membrane target, thereby reducing their transforming ability. It should be noted that levels of pp60^{src} kinase activity are unaffected by the amino-terminal mutations in rASV 1702 and rASV 157 (KARESS and HANAFUSA 1981). These viruses should provide unique tools to analyze the degree to which membrane phenomena are important in determining whether an in vivo neoplastic growth will be benign or malignant.

4.2 Temperature-Sensitive Mutant-Infected CEFs

CEFs infected with *ts*NY68 are temperature-sensitive for transformation: cells grown at permissive temperatures (33°–37 °C) have a rounded, transformed morphology and grow in soft agar colonies, while cells grown at restrictive temperatures (41°–42.5°) have a flat, untrasformed appearance and do not form colonies in soft agar (KAWAI and HANAFUSA 1971). Transformation of *ts*NY68-infected CEFs grown at 33° can be reversed if the cultures are shifted to 42°, and conversely, cells maintained at a restrictive temperature become retransformed when shifted to a permissive temperature. This cell system is ideal for study of molecular and cellular events occurring specifically in transformed cells and also permits study of these events during the transition period from an untransformed to a transformed state. Recently several investigators have found that the subcellar localization of pp60^{src} is temperature sensitive in *ts*NY68-infected CEFs (COURTNEIDGE and BISHOP 1982; BRUGGE et al. 1983; GARBER et al. 1983a). In *ts*NY68-infected cells grown at 33° or 35°, pp60^{src} fractionates chiefly with cellular membranes prepared by differential centrifugation of cell homogeneates. The degree of membrane association is comparable to that of pp60^{src} from CEF cultures infected with wild-type RSV. However, in *ts*NY68-infected cells grown at 41° or 42°, pp60^{src} fractionates largely as a soluble, cytosolic protein, whereas wild-type RSV-encoded pp60^{src} remains membrane-bound in cultures grown at 42°. Localization of pp60^{src} by immunofluorescence microscopy in *ts*NY68-infected cells (ROHRSCHNEIDER 1980; ROHRSCHNEIDER et al. 1982; GARBER et al. 1983a) suggests that pp60^{src} is associated with adhesion plaques and plasma membranes in cells grown at 33°. The fluorescence pattern of pp60^{src} in *ts*NY68-infected cells grown at 42° is diffuse, and neither adhesion plaque nor membrane staining is observed (Fig. 9). Thus *ts*NY68 pp60^{src} shows reduced interaction with cellular membranes when cells are grown at 42°. The failure of pp60^{src} to interact properly with membrane targets at 42° might play some role in reversing cellular transformation in these cells. It should be noted that pp60^{src} kinase activity also appears to be temperature sensitive in *ts*NY68 cells (COLLETT and ERIKSON 1978; LEVINSON et al. 1978). Thus the reversal of cell transformation in *ts*NY68 CEFs grown at 42° may be due to decreased phosphorylation of cell proteins, to decreased interaction of pp60^{src} with cellular membranes, or to a combination of both factors. An in vitro phosphorylation system which GARBER et al. (1983a) developed indicated that membranes prepared from RSV-transformed cells have much higher levels of tyrosyl kinase activity than membranes prepared from untransformed CEFs. Membrane proteins of 130 kd, 90 kd, and 68 kd are phosphorylated on tyrosine in a transformation-specific manner in membranes prepared from RSV-transformed cells. The same differences in levels of tyrosyl kinase activity between normal and transformed cells also was seen in membranes prepared from *ts*NY68-infected cells grown at 42° and 33° respectively. When *ts*NY68-infected cells were shifted from 42° to 33°, pp60^{src} rapidly reassociated with cellular membranes. The kinetics of this reassociation paralleled increases in tyrosyl kinase activity and phosphorylation of specific membrane proteins on tyrosine sites. A logical interpretation of these data is that pp60^{src} needs to be intimately associated with cellular membranes to directly or indirectly elicit phosphorylation of specific membrane proteins. Thus, both temperature-sensitive membrane association and temperature-sensitive kinase activity of *ts*NY68 pp60^{src} may contribute to its temperature-dependent transforming ability.

It is of interest that rASV 1702-infected cells appear morphologically transformed in comparison with *ts*NY68-infected cells grown at 42°, even though pp60^{src}s from both cells behave as soluble cytosolic proteins. This result might suggest that membrane association per se is relatively unimportant in causing morphological transformation of cells. However, pp60^{src} in rASV 1702-infected cells that had been fixed showed more prominent membrane staining than did pp60^{src} in *ts*NY68-infected cells maintained at 42 °C. Also "soluble" pp60^{src} in *ts*NY68 CEF grown at 42° is restricted within the cytosol by tight binding to the pp50:pp90 complex (see Sect. 7), whereas "soluble" pp60^{src}s encoded by rASV 1702 and rASV 157 appear to interact only transiently with this complex (KRUEGER et al., to be published). The role of membrane association of pp60^{src} in producing cell transformation may be further clarified by studying independent temperature-sensitive transformation mutants in the kinase and membrane-binding domains.

4.3 RR1022 Rat Cells, Pcl Goat Cells, and RSV-RR and RSV-Pcl Viruses in CEFs

RR1022 cells were derived from an in vivo sarcoma induced in an Amsterdam rat by SR-RSV-D. This cell line grows rapidly in tissue culture and efficiently forms colonies in soft agar. pp60^{src} was localized in these cells by immunofluorescence microscopy (KRUEGER et al. 1980b) and was associated both with the nuclear envelope membrane and with juxtanuclear reticular structures. It was not observed in plasma membranes or at regions of cell-to-cell contact. Cell fractionation studies showed that pp60^{src} was associated with cellular membranes and required detergent-containing

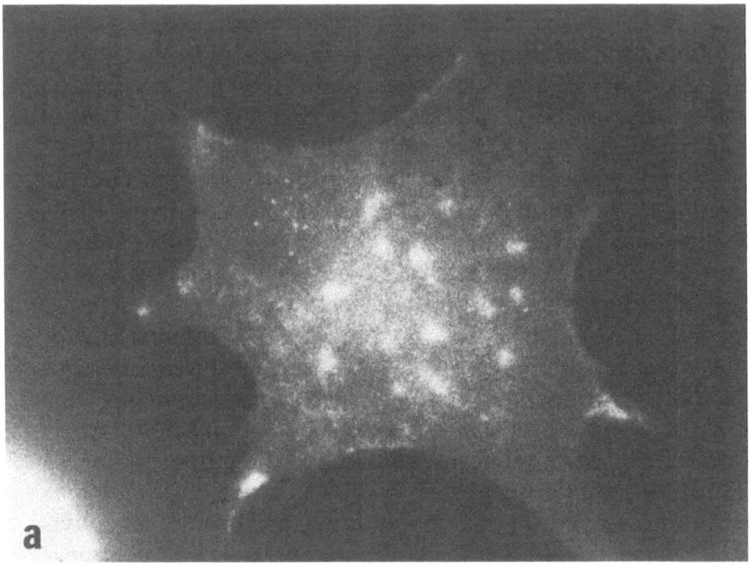

33°

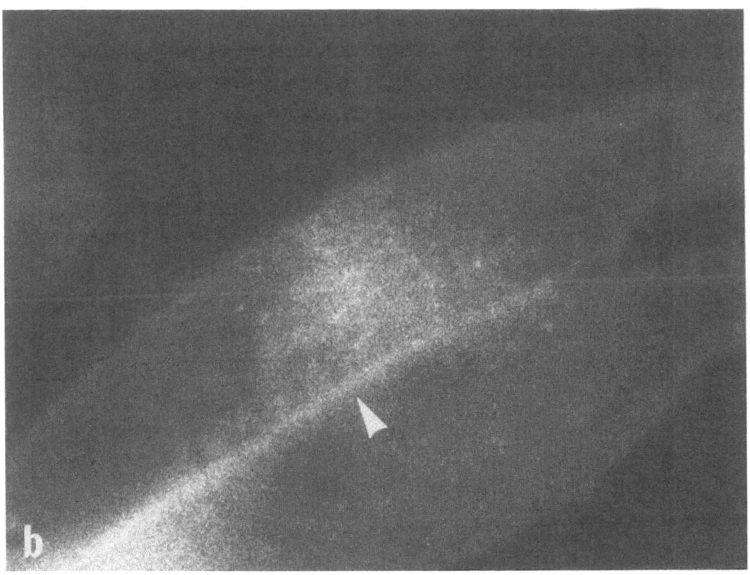

33°

Fig. 9a–d. Localization of $pp60^{src}$ in *ts*NY68-infected CEF by indirect immunofluorescence microscopy. In cells maintained at the permissive temperature (33°), $pp60^{src}$ can be seen in the free cell edge (**a, b**), at regions of cell-to-cell contact (**b,** *arrow*), in adhesion plaques (indicated by arrow in **c** and also visible as subnuclear speckles in **a**). When these cells are shifted to nonpermissive conditions (42°), $pp60^{src}$ is no longer associated with adhesion plaques or membranes, as indicated by the diffuse pattern of fluorescence (**d**)

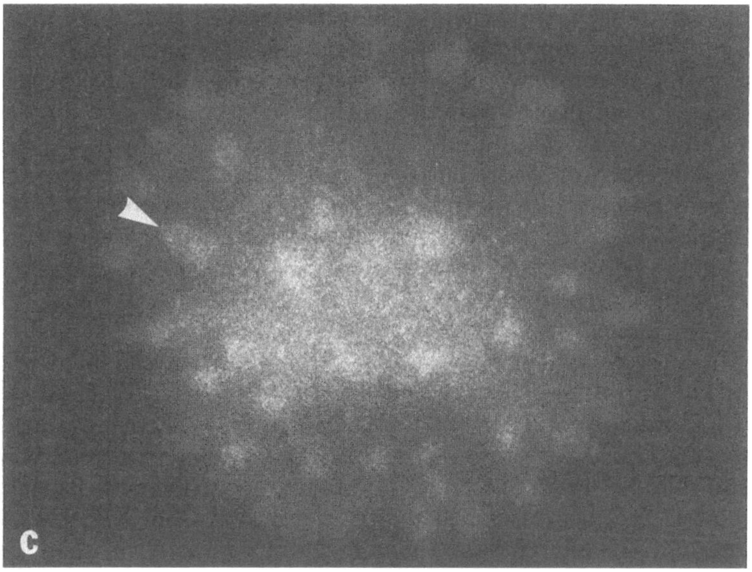

33°

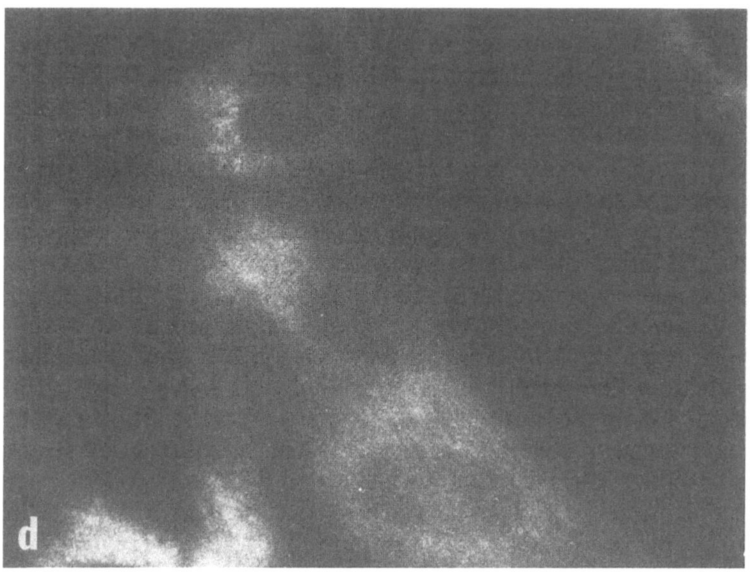

42°

Fig. 9c, d

buffers to release it from membranes. Because cell fractionation established that pp60^{src} was not a soluble cytoplasmic protein in RR1022 cells, the cytoplasmic fluorescence observed in these cells must have been the result of association of pp60^{src} with cytoplasmic membranes. The RR1022 pp60^{src} thus behaved like the integral membrane protein pp60^{src} in RSV-infected CEF, but appeared dissimilar in that it lacked certain amino-terminal *S. aureus* V8 protease sites and seemed to associate predominantly with different intracellular membranes. The novel localization of pp60^{src} in RR1022 cells might be due to its altered primary sequence (as suggested by the loss of protease-sensitive sites) or to altered processing of pp60^{src} in RR1022 cells.

The subcellular location of pp60^{src} was also investigated in Pcl cells, goat skin fibroblasts transformed in vitro with SR-RSV-D (Garber et al. 1982). Immunofluorescence staining of pp60^{src} in Pcl cells suggested that it was associated with internal cellular membranes in a fashion similar to that in RR1022 cells. Since pp60^{src} isolated from Pcl cells lacked the same amino-terminal protease sites as pp60^{src} isolated from RR1022 cells, it appeared that the altered subcellular location of pp60^{src} in these cells might be best explained by an altered structure of pp60^{src}. To test this hypothesis, transforming viruses (called RSV-RR and RSV-Pcl) were rescued from both RR1022 cells and Pcl cells, and CEF cells were infected with these viruses (Garber et al. 1982). In this way, the localization of RSV-RR and RSV-Pcl pp60^{srs} could be compared with that of wild-type pp60^{src} in the same cell type. The structural change in RR1022 and Pcl pp60^{src}s which accounts for the loss of amino-terminal *S. aureus* protease sites was retained in RSV-RR- and RSV-Pcl-encoded pp60^{src}s in CEF cultures. The immunofluorescence pattern of pp60^{src} in CEF cells transformed with these viruses suggests that pp60^{src} is associated predominantly with nuclear envelope and internal cytoplasmic membranes. Immunofluorescence staining of endoplasmic reticulum and other intracellular membranes with antibodies to cytochrome P-450 reductase showed a similar pattern to pp60^{src} fluorescence in these cells. Differential centrifugation of homogenates of RSV-RR- and RSV-Pcl-transformed CEF demonstrated that more pp60^{src} was associated with nuclear fractions than pp60^{src} encoded by wild-type RSV (Table 5). Since pp60^{src} is released from these nuclear fractions by detergents, it is probably associated with nuclear envelope and perinuclear membraneous structures. The data in Table 5 establish that RSV-RR and RSV-Pcl pp60^{src}s are not soluble cytosolic proteins and are associated predominantly with cellular membranes. This result confirms the immunofluorescence localization of pp60^{src} to cytoplasmic membranes in these cells. More detailed information about the association of pp60^{src} with specific intracellular membranes in RSV-RR- and RSV-Pcl-infected CEFs was obtained by membrane fractionation in isopyknic sucrose gradients (Table 6). The fractionation of subcellular organelles in these gradients suggests that RSV-RR and RSV-Pcl pp60^{src}s are more highly associated with endoplasmic reticulum than plasma membranes. A minority of intracellular pp60^{src} may associate with plasma mem-

Table 5. Distribution of $pp60^{src}$ kinase activity (%) in subcellular fractions from RSV-transformed cells fractionated by differential centrifugation

Fraction	Cell type				
	RSV-D CEF	RSV-RR CEF	RSV-Pcl CEF	RR1022	Pcl
N	10	32	26	68	57
P_{10}	36	42	42	15	24
P_{100}	21	14	16	3	4
S_{100}	33	12	16	15	14

Table 6. Distribution of $pp60^{src}$ kinase activity (%) in subcellular fractions from RSV-transformed cells fractionated by isopyknic centrifugation

Fraction	Cell type				
	RSV-D CEF	RSV-RR CEF	RSV-Pcl CEF	RR1022	Pcl
Plasma membrane: gradient fractions 1 and 2	54	28	30	25	27
Endoplasmic reticulum (smooth > rough): fraction 3	28	17	22	8	12
Endoplasmic reticulum (rough > smooth): fraction 4	12	44	38	46	31
Mitochondira and ribosome: fraction 5	6	11	10	21	30

branes in RSV-RR- and RSV-Pcl-transformed CEF, since plasma membrane fractions do contain some $pp60^{src}$ and since the distribution of $pp60^{src}$ in these gradients is bimodal. The reduced amount of RSV-RR $pp60^{src}$ in plasma membranes does not appear to alter its ability to induce malignant, anaplastic sarcomas in chickens (KRUEGER, unpublished work). However, the subcellular localization of $pp60^{src}$ in sarcoma cells in vivo has not been investigated. Since a lesser quantity of $pp60^{src}$ in association with plasma membranes may be sufficient to induce cellular transformation, these data do not strongly disfavor the hypothesis that the plasma membrane is an important target for the transforming action of $pp60^{src}$. In summary, $pp60^{src}$s encoded by RSV-RR and RSV-Pcl appear to be integral membrane proteins which show reduced plasma membrane association. It is likely that structural changes in these $pp60^{src}$s cause the altered subcellular distribution. Similar alterations in $pp60^{src}$ structure and subcellular location could occur in other types of mammalian cells transformed by SR-RSV-D.

4.4 CU Mutant-Infected CEFs

In order to dissect the mechanism of pp60^{src}-mediated transformation, WEBER and co-workers have isolated a number of partial transformation mutants (ANDERSON et al. 1981). Immunofluorescence microscopy has been used to localize pp60^{src} in three partial transformation mutants: CU2, CU12, and *ts*CU11 (ROHRSCHNEIDER et al. 1982; ROHRSCHNEIDER and ROSOK 1983). These data deal principally with the localization of pp60^{src} in adhesion plaques and are discussed in Sect. 6.

5 Lipid Is Tightly Associated with pp60^{src}

The models of pp60^{src} (Fig. 7a, b) propose that pp60^{src} interacts with membranes via an amino-terminal hydrophobic domain. However, the predicted amino acid sequence of pp60^{src} contains no long uninterrupted stretch of hydrophobic residues within the amino-terminus (TAKEYA and HANAFUSA 1982; TAKEYA et al. 1982). The association of most *src*-specific mRNA with free polysomes (LEE et al. 1979; PURCHIO et al. 1980) and pulse-chase studies of pp60^{src} synthesis (LEVINSON et al. 1981) suggest that pp60^{src} is synthesized as a soluble protein that later becomes membrane-associated through a post-translational modification. Recently, it has been demonstrated that fatty acid is tightly associated with some membrane-bound proteins (SCHMIDT and SCHLESINGER 1979, 1980; SCHMIDT et al. 1979; SCHLESINGER et al. 1980; SCHMIDT 1982). To test for fatty acid binding to pp60^{src}, SEFTON et al. (1982b) added [^{3}H]palmitate to cultures of RSV-transformed cells to label phospholipids and lipoproteins. Immunoprecipitation of cell extracts with TBR serum followed by gel electrophoresis of the immunoprecipitated material identified a 60-kd lipid-labeled protein which comigrated with pp60^{src}. This protein was identified as pp60^{src} by limited proteolytic mapping with *S. aureus* V8 protease. The ^{3}H label was found in amino-terminal 34-kd and 18-kd peptides of pp60^{src}, and was recovered as [^{3}H]palmitate when the immunoprecipitated protein was hydrolyzed in acid and the hydrolysate was analyzed by thin layer chromatography. These data suggest that the added [^{3}H]palmitate was not degraded and used to synthesize labeled amino acids and that the palmitate was bound to pp60^{src} somewhere within 18 kd of the amino-terminus of pp60^{src}. Since the palmitate remains attached to pp60^{src} after boiling in SDS, it is likely that it is covalently attached to pp60^{src}. The attachment is probably not through an ester linkage, since it is stable in alkaline solutions which hydrolyze ester bonds. Recently, amino-terminal myristic acid amide linkages have been described in bovine cardiac muscle cyclic AMP-dependent protein kinase (CARR et al. 1982) and in mammalian type B, C, and D retroviral *gag* proteins (HENDERSON et al. 1983; SCHULTZ and OROSZLAN 1983). It is possible that myristic acid, an intermediate in the biosynthesis of palmitic acid, may be the lipid moiety bound to pp60^{src}.

5.1 Lipid Binding to pp60^{src} Correlates with Membrane Binding and Integration

GARBER et al. (1983b) studied the functional significance of lipid binding to pp60^{src}. rASV 1702 and rASV 157 have deleted or added sequences within the amino-terminal 8-kd membrane-binding domain of pp60^{src} (see Sect. 3.2, 4.1). CEF cultures transformed by these rASVs and control rASVs encoding 60-kd pp60^{src}s were labeled with [^{3}H]palmitate and pp60^{src} was immunoprecipitated from these cells. Membrane-associated pp60^{src}s encoded by RSV and rASV 1441 contained equivalent amounts of [^{3}H]palmitate. The size-variant pp60^{src}s encoded by rASV 1702 and rASV 157, which are not integrated into cellular membranes, did not contain detectable amounts of [^{3}H]palmitate. Also, pp52^{src} (derived from RSV-encoded pp60^{src} by removal of the amino-terminal 8 kd) did not contain detectable [^{3}H]palmitate. The palmitate labeling of pp60^{src} is shown in Fig. 10. These data suggest that fatty acid is attached to pp60^{src} within the amino-terminal membrane-binding domain. As discussed in Sect. 3.3, the amino acid sequence predicted for the membrane-binding domain does not show long uninterrupted stretches of hydrophobic amino acids. It seems likely that the hydrophobic properties of this region may be explained by a combination of attached lipid and smaller regions of hydrophobic amino acids.

As mentioned in Sect. 4.2, the intracellular localization of pp60^{src} encoded by *ts*NY68 is temperature dependent. The diminished ability of *ts*NY68 pp60^{src} to transform cells at 42° presumably reflects some denaturation of region(s) of the molecule, thereby rendering it inactive. The membrane-binding properties of proteins attached to membranes through hydrophobic interactions should not be affected by slight protein denaturation (see Table 4). It was somewhat surprising that *ts*NY68 pp60^{src} was removed from plasma membranes by incubation of the cells at 41°–42° (COURTNEIDGE and BISHOP 1982; GARBER et al. 1983a). In an effort to explain this temperature-sensitive redistribution of pp60^{src}, the amount of [^{3}H]palmitate bound to *ts*NY68 pp60^{src} at 33° and 42° was measured (GARBER et al. 1983b). Membrane-associated pp60^{src} from cells grown at 33° contained a similar amount of [^{3}H]palmitate as membrane-associated pp60^{src} encoded by RSV. Both membrane binding of pp60^{src} and the amount of lipid bound to pp60^{src} were reduced three- to fourfold in *ts*NY68-infected cells grown at 42°. These data suggest that hydrophobicity of the amino-terminal membrane-binding region may be increased by lipid attachment and that this attachment may direct the integration of pp60^{src} into membranes.

Another functional significance of lipid binding is suggested by the experiments of ITO et al. (1982). Anionic phospholipids were found to stimulate the phosphorylation of vinculin, a putative pp60^{src} substrate (SEFTON et al. 1981), and inhibit the phosphorylation of casein and other nonphysiological substrates by a 54-kd *src* tyrosyl kinase purified from RSV-induced tumors in rats (BLITHE et al. 1982). Interestingly, they found that palmitic acid did not stimulate vinculin phosphorylation. It is not known whether the

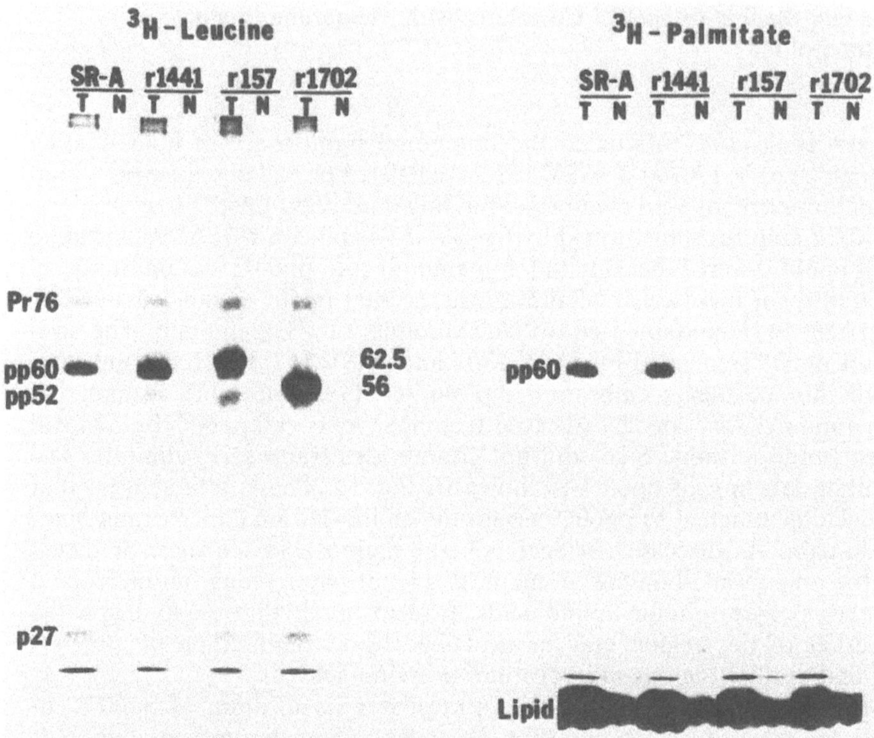

Fig. 10. Association of lipid with pp60^{src}. CEFs were transformed by infection with SR-RSV-A (*SR-A*), rASV 1441, rASV 157, or rASV 1702. Cells were labeled metabolically with either ^{3}H-leucine or ^{3}H-palmitic acid and lysed with a detergent-containing buffer. Lysates were immunoprecipitated either with TBR (*T*) or nonimmune (*N*) sera. Proteins in the immunoprecipitates were analyzed by SDS-PAGE

54-kd pp60^{src} fragment retains the lipid binding site, but their results suggest that lipid can enhance the action of pp60^{src}.

5.2 Revised Model of pp60^{src} Interaction with Membranes

We used the information contained in Fig. 7a, b, the data shown in Fig. 10, and the sequence data of TAKEYA et al. (1982), to construct a new model for the interaction of pp60^{src} with cellular membranes. This model, shown in Fig. 11, depicts pp60^{src} as an integral membrane protein which associates with the periplasmic face of plasma membranes (or other intracellular membranes in the case of some mutants). pp60^{src} is attached to this membrane by a 8-kd amino-terminal hydrophobic membrane-anchoring domain. The hydrophobicity of this domain is due, at least in part, to attached fatty acid. It is unlikely that hydrophobic amino acids located outside this 8-kd domain contribute to membrane binding, since deletion of hydrophobic

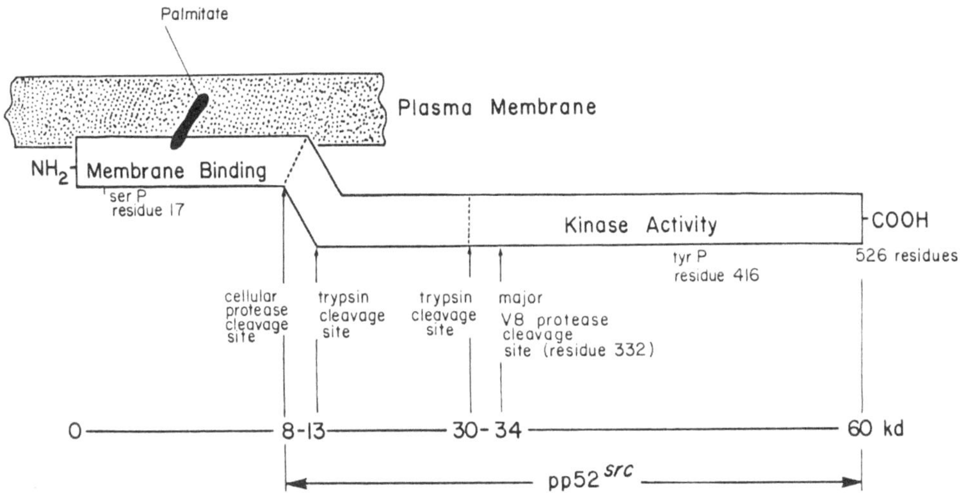

Fig. 11. Schematic model of pp60^{src} indicating structural and functional domains. Palmitate is bound to pp60^{src} within the 8-kd amino-terminal membrane-binding region (other types of lipid could also be bound within this region). Several protease sites which were defined in Fig. 7 are indicated. The model indicates that pp60^{src} is an integral, monotopic membrane protein which binds to membranes through hydrophobic interactions. The indicated size of the membrane-binding domain is a maximum estimate, and a smaller portion of this region could be integrated into the plasma membrane

amino acids located outside this domain does not affect membrane binding (F. Cross, personal communication). The kinase activity of pp60^{src} is restricted to its carboxy-terminal half. The locations of phosphoserine and phosphotyrosine within pp60^{src} are indicated.

6 Localization of pp60^{src} in Adhesion Plaques

As mentioned previously, pp60^{src} has been observed in adhesion plaques in RSV-transformed cells. Since adhesion plaques mediate cell substratum attachment and serve as anchor points for cytoskeleton insertion into membranes, the interaction of pp60^{src} with adhesion plaques could explain many diverse properties of transformed cells.

6.1 Introduction to Cell-Substratum Adhesions

Adhesion plaques or "cell feet" are points of attachment between cell membranes and the surface on which the cells are grown in vitro cell culture systems. These structures were first described by Curtis (1964) who used interference reflection microscopy to examine cell-substratum interaction.

These studies suggested that some areas of the ventral cell surface were located within approximately 10 nm of the glass substratum on which the cells were grown. Examination of cultured cells by electron microscopy confirmed the existence of discrete regions of cell-substratum interaction where the ventral cell membrane was separated from the substratum by only 10–15 nm (ABERCROMBIE et al. 1971). In contrast, regions of the ventral cell surface apparently uninvolved in cell atachment were located up to 100 nm from the substratum. Subsequently, three types of cell-substratum interaction have been defined by interference reflection microscopy in cultured chick fibroblasts (IZZARD and LOCHNER 1976):

1. Focal contacts are 0.25 μm wide × 0.5 μm long cell adhesions which come within 10–15 nm of the substrate and appear as dark or black streaks in the interference reflection microscope.
2. Close contacts are more extensive, patch-like regions where the cell surface is separated by about 30 nm from the substratum. Close contacts form predominantly towards the leading cell edge in migrating fibroblasts and are recognized by an "iron grey" or "grey fringe" interference color.
3. Regions of separation are regions where the ventral cell surface membrane is separated by approximately 100–150 nm from the culture surface and appear as white areas in the interference reflection microscope.

Examination of chick fibroblasts by both interference reflection and high-voltage electron microscopy demonstrated that focal contacts were always associated with the distal end of microfilament bundles, whereas close contacts were associated with a meshwork of microfilaments (HEATH and DUNN 1978). The intimate association of microfilament bundle termini and focal adhesions also was seen clearly in electron micrographs of replicas of the ventral cell surface of cultured cells (REVEL and WOLKEN 1973). Studies using interference reflection and immunofluorescence microscopy have confirmed that the distal end of microfilament bundles (stress fibers) terminate in focal adhesions (WEHLAND et al. 1979).

Focal adhesions, which also have been termed cell feet, and adhesion plaques, have been shown to contain numerous cytoskeletal proteins, including actin (WEHLAND et al. 1979), alpha-actinin (WEHLAND et al. 1979), vinculin (GEIGER, 1979), filamin (ROHRSCHNEIDER and ROSOK 1983), and gelsolin (WANG et al. 1983). Focal adhesions seem to provide a direct transmembrane linkage between intracellular cytoskeleton elements and the extracellular fibronectin network. In electron microscopic studies, SINGER (1979) observed a close association (maximum separation 8–22 nm) between actin microfilaments and fibronectin cables. In many instances microfilament termini seemed to be linked (in an end-to-end fashion) with fibronectin fibers through a dense plasma membrane plaque which appears similar to focal contacts observed in other studies. Singer proposed the term fibronexus to indicate the possible function of the microfilament and fibronectin as a unit. Immunofluorescence and interference reflection microscopy show a close correlation between actin microfilaments, vinculin in focal adhesions, and pericellular fibronectin fibers (HYNES and DESTREE 1978; SINGER and

PARADISO 1981) thus supporting the linkage of fibronectin to actin microfilaments via focal contacts. While the fibronexus is undoubtedly a key feature of cell-substratum cohesiveness, recent evidence suggests that cell attachment is independent of the fibronectin-adhesion plaque interaction (BIRCHMEIER et al. 1982). Presumptive transmembrane proteins of the focal adhesion which mediate its interaction with the substratum and with the fibronexus are unknown.

Focal adhesions thus seem reasonably well characterized as structural elements of cell attachment, but their possible functional significance in controlling dynamic cellular processes is only poorly understood. In this regard, focal adhesions are transient structures in migrating fibroblasts which form in the direction of cell movement and disassemble in the trailing cell edge, and thus appear to mediate or control cell locomotion (GEIGER 1982). Migrating chick heart fibroblasts in culture form a type of cell-to-cell junction which has a 10-nm separation between opposing plasma membranes and appears similar to the focal adhesion in the electron microscope (HEAYSMAN and PEGRUM 1973). These intercellular junctions form within 20 s of contact and may mediate "contact inhibition" in these cells. Focal contacts may also be involved in controlling "contact guidance" of cultured cells (DUNN and HEATH 1976). Thus, cells are in direct contact with their environment through focal adhesions, and these structures may influence more dynamic aspects of adherence, migration, and growth control of cultured cells. There is no strict equivalent of the adhesion plaque in more highly ordered tissues in vivo, but adherens-type junctions which occur in some tissues (e.g., zonula adherens of intestine epithelium, fascia adherens of heart myocardium, and dense plaques of smooth muscle) are points of actin microfilament insertion into membranes and contain both alpha-actinin and vinculin (GEIGER et al. 1980). These adherens-type junctions clearly serve a structural function as anchor points for actin microfilaments, but any role for them in controlling dynamic cellular processes in vivo is unknown.

6.2 Adhesion Plaques in RSV-Transformed Cells

In the discussion below, "adhesion plaque" refers to regions of cell-substratum contact which occur in transformed cells and are visible as black or grey focal adhesions on interference reflection microscopy. Cell-substratum adhesions in RSV-transformed rat kidney cells are characterized by (a) decreased focal adhesions and close contacts, and (b) the appearance of a new type of cell-substratum contact which may be smaller or larger than a focal adhesion and has the interference reflection color of the close contact (DAVID-PFEUTY and SINGER 1980). The term adhesion plaque thus encompasses the "classic" focal adhesion of IZZARD and LOCHNER (1976) and the type of altered cell-substratum contact described by DAVID-PFEUTY and SINGER (1980) in RSV-transformed cells. Using immunofluorescence microscopy, ROHRSCHNEIDER (1980) localized pp60^{src} within adhesion plaques in

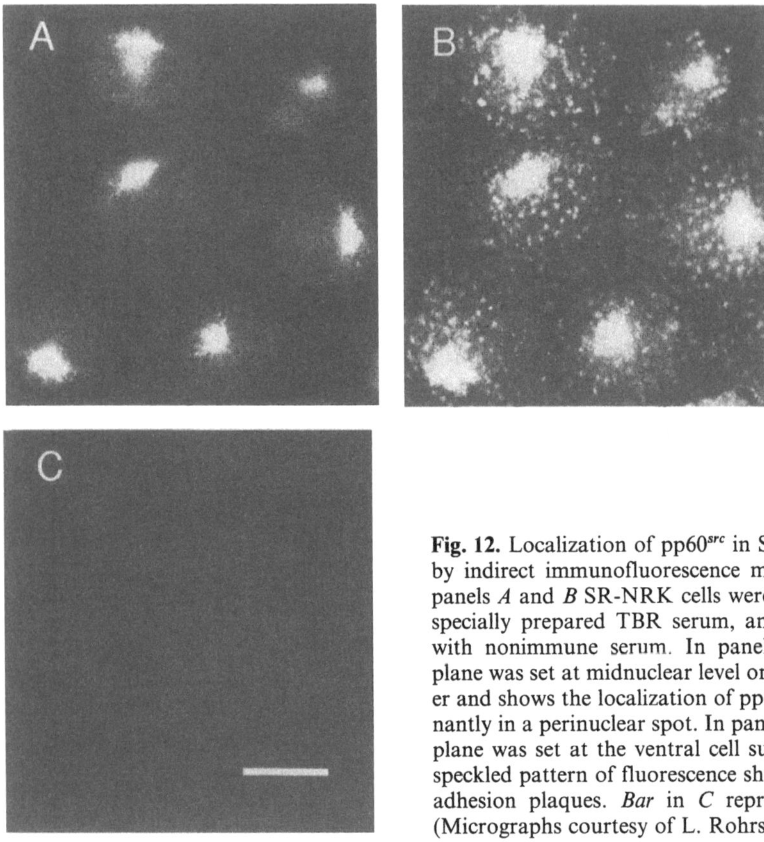

Fig. 12. Localization of pp60^{src} in SR-NRK cells by indirect immunofluorescence microscopy. In panels *A* and *B* SR-NRK cells were reacted with specially prepared TBR serum, and in panel *C* with nonimmune serum. In panel *A*, the focal plane was set at midnuclear level or slightly higher and shows the localization of pp60^{src} predominantly in a perinuclear spot. In panel *B*, the focal plane was set at the ventral cell surface and the speckled pattern of fluorescence shows pp60^{src} in adhesion plaques. *Bar* in *C* represents 20 μm. (Micrographs courtesy of L. Rohrschneider)

SR-NRK cells, RSV-transformed CEFs, RSV-transformed mouse fibroblasts, and *ts*NY68-infected CEFs grown under permissive conditions. pp60^{src} was not present in adhesion plaques of *ts*NY68-infected CEF grown under restrictive conditions. In these studies interference reflection microscopy was used to verify that the fluorescence pattern of pp60^{src} coincided with adhesion plaques. The localization of pp60^{src} in adhesion plaques of SR-NRK cells is very striking and is shown in Fig. 12.

SR-NRK cells were used in subsequent experiments designed to characterize the structure of adhesion plaques and the interaction of pp60^{src} within these structures (Shriver and Rohrschneider 1981 a, b). In sparse cultures, adhesion plaques containing pp60^{src} were organized prominently in a sub- and perinuclear location, with more diffuse adhesion plaques occurring in the cytoplasm and cell periphery. In more confluent cultures, pp60^{src} was found principally in adhesion junctions, a type of cell-to-cell contact observed in transformed SR-NRK cells but not in untransformed NRK cells. Within adhesion junctions pp60^{src} stained in the midline between cells, suggesting that it was concentrated on the inner face of the plasma membrane at these regions of cell-to-cell contact. Actin, alpha-actinin, and vinculin also were contained in adhesion junctions, but were excluded from the midline, suggesting a slightly different (perhaps more cytoplasmic) location than $pp60^{src}$ within this structure. Additional evidence that $pp60^{src}$ is located in adhesion plaques comes from immunofluorescence microscopy using antibody directed against a carboxy-terminal peptide sequence of $pp60^{src}$ (NIGG et al. 1982a). In these experiments, antibodies raised against a synthetic peptide were affinity-purified. In this way, it was possible to prepare monospecific antisera to a particular antigenic site on $pp60^{src}$. These monospecific antibodies showed both adhesion plaque and plasma membrane staining in NRK cells transformed by the B77 strain of RSV. Since plasma membrane staining also was found in SR-NRK cells with anti-$pp60^{src}$ antibodies, one may conclude that some, but not all, intracellular $pp60^{src}$ is organized in adhesion plaques in RSV-transformed rat kidney cells.

In order to quantitate the amount of $pp60^{src}$ contained in adhesion plaques, ROHRSCHNEIDER et al. (1982) isolated adhesion plaques from [^{35}S]methionine-labeled SR-NRK cells using a combination of techniques that included detergent extraction and mechanical shearing. The amount of $pp60^{src}$ that could be immunoprecipitated from these isolated adhesion plaques was 12% of the total amount of immunoprecipitable $pp60^{src}$ from unfractionated cells. $pp60^{src}$ appeared to be fivefold concentrated in adhesion plaques with respect to its total cellular concentration. It could be identified directly in adhesion plaques by incubation of isolated plaques in buffer containing gamma-labeled [^{32}P]ATP and Mg^{2+}. Under these conditions, $pp60^{src}$ was phosphorylated on tyrosine and was identified by two-dimensional gel electrophoresis of proteins solubilized from adhesion plaques. Interestingly, the normal cellular *src* protein, $pp60^{C\text{-}src}$, also could be phosphorylated in adhesion plaques isolated from untransformed NRK cells. It is not clear whether $pp60^{src}$ autophosphorylates or is phosphorylated by another tyrosyl kinase present in the adhesion plaque preparation.

The organization of $pp60^{src}$ within adhesion plaques was investigated in cells infected with partial transformation mutants (ROHRSCHNEIDER et al. 1982; ROHRSCHNEIDER and ROSOK 1983). CEF cultures infected with the mutants CU2, CU12, and *ts*CU11 showed loss of some parameters of cell transformation while retaining others (ANDERSON et al. 1981). These mutants showed quite different cellular morphologies and differed in their ability to produce cellular transformation in vitro and in vivo (see Table 7). Immunofluorescence staining with antivinculin antibodies also showed that the organization of adhesion plaques is quite different in CEF cultures infected with these mutants (ROHRSCHNEIDER et al. 1982). The presence of $pp60^{src}$ in adhesion plaques in the CU-mutant-infected cells is listed in Table 7. Colony formation in soft agar is independent of the presence of $pp60^{src}$ in adhesion plaques, since CU12-transformed CEFs form very large colonies in soft agar but do not contain detectable levels of $pp60^{src}$ in adhesion plaques. Vinculin phosphorylation on tyrosine is also independent of the presence of $pp60^{src}$ in adhesion plaques: CU2-infected CEFs have abundant adhesion plaques containing $pp60^{src}$ but low levels of phosphotyrosine in vinculin, whereas $pp60^{src}$ is not detected in adhesion plaques of

Table 7. Localization of pp60^{src} in CEFs infected with different viruses and some properties of these viruses and transformed cells

Virus	Transformed cell morphology	Soft agar colony formation	Tumorigenicity of virus in chickens	Subcellular location of pp60 by biochemical cellular fractionation	Adhesion plaque abundance: from actin, vinculin, or filamin staining and interference reflection microscopy	Adhesion plaque arrangement: presence of abnormal aggregates and clusters	Abundance of pp60 in adhesion plaques	Increased vinculin phosphorylation on tyrosine	Actin stress fibers	Cell surface fibronectin
SR-RSV-A	Round, refractile	+	High	Plasma membrane	+	–	– to +[a]	+	–	–
*ts*NY68 37°	Round, refractile	+		Plasma membrane	+	–	– to +[a]	+	–	–
*ts*NY68 42°	Normal	–		Soluble cytosol	++	–	–	–	+	+
CU2	Flat, blebby	+	Reduced	ND	++	++	++	–	+	–
CU12	Fusiform	+		ND	+	–	–	+	+	+
*ts*CU11 37°	Flat, well spread, similar to same cells at 42°	+		ND	+	–	+	+	–	–
*ts*CU11 42°	Normal (flat, well spread)	–		ND	++	–	–	–	+	+
rASV 1702	Spindle-formed to moderately well spread	+	Reduced	Soluble cytosol (salt sensitive)	++	+++	++	ND	+[b]	+
rASV 157	Spindle-formed to moderately well spread	+	Reduced	Soluble cytosol (salt sensitive)	++	+++	++	ND	+[b]	+
rASV 3811 or rASV 1441	Round to spindle-form	+	High	Plasma membrane	+	–	+	ND	+[b]	–
RSV-RR	Extremely round, refractile	+	High	Membrane (endoplasmic reticulum, plasma membrane)	ND	ND	ND	–	–	ND
RD10	Flat	+	High	ND	++	+	++	+	+	–

[a] Usually pp60^{src} is seen only in larger cells which have more extensive adhesion plaques

[b] Some spread cells have a few actin stress fibers which can be seen with NBD-phallacidin staining, but all cells have fewer actin cables than *ts*NY68 at 42° or uninfected CEF

CU12-transformed CEFs although these cells contain vinculin with high levels of phosphotyrosine.

The organization of pp60^{src} within adhesion plaques of rASV-transformed CEFs also has been investigated (KRUEGER et al., to be published). CEFs transformed by RSV show decreased focal contacts and close contacts, but most remaining focal contacts appear streak-like. In contrast, CEFs transformed by rASV 1702 and rASV 157 (which encode size-variant, "soluble" pp60^{src}s) produce large aggregates of round adhesion areas which appear gray in the interference reflection microscope and are therefore probably not as closely adherent to the substratum as focal contacts in uninfected cells or those remaining in RSV-transformed CEFs. Cells transformed by control rASV 3811 or rASV 1441 (which encode normal-sized, membrane-integrated pp60^{src}s) show a mixture of normal focal contacts and smaller, punctate adhesions with a gray interference color. The adhesion plaques in rASV 1702- and rASV 157-transformed cells thus appear to be "overproduced", as in CU2-infected CEFs or in SR-NRK cells (the overproduction referring more to the size and arrangement of the adhesion areas than to the total surface area occupied by adhesions). The pp60^{src}s encoded by rASV 1702 and rASV157 fractionate as soluble, cytosolic proteins in a salt-sensitive fashion, indicating that they are not integral membrane proteins. However, as determined by immunofluorescence microscopy, these *src* proteins associate with the large adhesion plaques in these cells (and show a limited ability to interact with plasma membranes at regions of cell-to-cell contact). These "soluble" pp60^{src}s thus fit the category of peripheral membrane proteins which interact with membrane regions via association with other membrane proteins or complexes (see Sect. 4.1). Consistent with this interaction, these "soluble" pp60^{src}s do not show a general membrane-staining pattern of pp60^{src} such as that seen in CEFs transformed by control rASV 3811. Thus, the integration of pp60^{src} into the plasma membrane appears to be separable from its association with adhesion plaques, though these two interactions are not necessarily exclusive in cells transformed by wild-type RSV.

Table 7 lists some properties of CEFs transformed by RSV, rASVs, and other RSV mutants. Included in this table are data relevant to pp60^{src} localization within adhesion plaques in these cell types and the subcellular localization of pp60^{src} as determined by biochemical cellular fractionation. Also listed is the ability of these viruses to produce sarcomas in chickens. These data indicate no correlation between localization of pp60^{src} in adhesion plaques and either cellular morphology or the ability of the transformed cells to grow in soft agar colonies. Cells infected with CU2, rASV 1702, or rASV 157 "overproduce" abnormal-looking clusters of adhesion plaques containing pp60^{src}, and interestingly, all of these viruses show extremely reduced ability to produce tumors in chickens. Conversely, cells transformed by viruses which show increased tumorigenicity have widely variable amounts of pp60^{src} associated with adhesion plaques. Vinculin phosphorylation on tyrosine is independent of the presence of pp60^{src} in adhesion plaques, but it is possible that vinculin phosphorylation can occur at multiple intracellular sites, since vinculin is found in both adhesion plaques and a detergent-soluble cytoplasmic compartment (GEIGER 1982). However, vinculin phosphorylation on tyrosine is not required for cell transformation, since CEFs infected with RSV-RR (see Sect. 4.3) appear morphologically transformed and the virus has a normal ability to produce tumors in vivo, but there is no increased phosphorylation of vinculin on tyrosine in these transformed cells (Table 7) (ANTLER et al. to be published). The presence of $pp60^{src}$ in adhesion plaques also fails to correlate with either stress fiber dissolution or with a decreased abundance of extracellular fibronectin (Table 7).

No clear functional role for $pp60^{src}$ in adhesion plaques is suggested by these data, but the variant $pp60^{src}$s described in Table 7 may have altered interaction with as yet unidentified structural proteins that modify adhesion plaque function. Insight into the biochemical interaction of $pp60^{src}$ with adhesion plaque constituents may come from further study of isolated adhesion plaques in in vitro systems or from study of additional RSV mutants. Since the localization by immunofluorescence of $pp60^{src}$ within adhesion plaques is so striking, it is tempting to speculate on the importance of this interaction. However, we urge caution in this respect since immunofluorescence microscopy yields only qualitative data. Furthermore, the majority of intracellular $pp60^{src}$ (e.g. >85% in SR-NRK cells) is not contained in adhesion plaques but rather is found associated with cellular membranes. It also should be noted that SR-NRK cells growing in soft agar colonies do not form adhesion plaques or specialized intercellular junctions (PASTAN et al. 1982). This provides additional evidence that $pp60^{src}$ does not need to be present in these structures to produce altered cell growth properties.

7 Transport of $pp60^{src}$ to the Membrane

By what mechanism is $pp60^{src}$ transported to the plasma membrane? It has been reported that a cell-free reticulocyte translation reaction mixture primed with RSV virion RNA results in the synthesis of a 62-kd $pp60^{src}$ precursor that can be specifically cleaved to 60 kd upon addition of canine pancreas membrane vesicles to the translation system (KAMINE and BUCHANAN 1978; KAMINE et al. 1980), suggesting that $pp60^{src}$ is an integral membrane protein synthesized as a precursor with a signal sequence. However, cell-free translation of 21S *src*-specific mRNA from RSV-transformed cells in the reticulocyte lysate system indicated that the primary translation product was $pp60^{src}$, with no evidence for processing of a signal sequence from a precursor (SEFTON et al. 1978; BISHOP et al. 1980). The nucleotide sequence of the *src* gene also provides no evidence for a hydrophobic signal sequence at the amino-terminus (TAKEYA et al. 1982). Furthermore, $pp60^{src}$ is synthesized in the cytosol on free polyribosomes (LEE et al. 1979; PURCHIO et al. 1980), suggesting that $pp60^{src}$ possesses no signal sequence and that membrane attachment occurs post-translationally. Pulse-chase studies indicate that $pp60^{src}$ is synthesized in the cytosol and becomes membrane associated within 10–15 min of synthesis (LEVINSON et al. 1980). Characterization of the interaction of pp60^{src} with two cellular phosphoproteins, pp50 and pp90, suggests that the pp50:pp60^{src}:pp90 complex may be involved in the processing of pp60^{src} before the association of pp60^{src} with the plasma membrane (COURTNEIDGE and BISHOP 1982; BRUGGE et al. 1983).

Both pp50 and pp90 coimmunoprecipitate with pp60^{src}; pp90 contains phosphoserine, and pp50 contains phosphoserine and phosphotyrosine (SEFTON et al. 1978; HUNTER and SEFTON 1980; OPPERMANN et al. 1981a; BRUGGE et al. 1981). They are not immunoprecipitated by TBR serum from uninfected or *td*RSV-infected cells, are not related to any RSV structural proteins, and are not structurally related to pp60^{src} (BRUGGE et al. 1981; OPPERMANN et al. 1981a). Sedimentation analysis of lysates from RSV-infected cells revealed that 5%–10% of the pp60^{src} sediments with pp50 and pp90 as a complex, while the remainder sediments as a monomer. In lysates from *ts*NY68- or *ts*LA90-infected cells grown at either the permissive or restrictive temperature, the proportion of bound pp60^{src} is increased. The complex dissociates in low salt, but is unaffected by high salt, disulfide bond-reducing agents, or divalent cation chelators. The bound form of pp60^{src} contains little or no stable phosphotyrosine and has little or no immune complex kinase activity, while the free form of pp60^{src} contains phosphotyrosine and has kinase activity (BRUGGE et al. 1981; COURTNEIDGE and BISHOP 1982; BRUGGE et al. 1983).

pp90 is a major cytoplasmic protein of normal avian cells whose synthesis is amplified in response to stress conditions such as heat shock, treatment with arsenite or canavanine, or glucose starvation (OPPERMANN et al. 1981b; BRUGGE et al. 1981, 1983). There has been a report that the antigenic determinants of pp90 that are reactive with rabbit antiserum to purified avian heat shock pp90 (chsp89) are sequestered in the pp50:pp60^{src}:pp90 complex (OPPERMANN et al. 1981b). However, BRUGGE et al. (1983) demonstrated that mouse monoclonal antibody to partially purified avian brain pp90 can immunoprecipitate the complex as well as free pp90. pp90 appears to be a highly conserved protein in widely divergent species, but does not appear to be a glycolytic enzyme (KELLEY and SCHLESINGER 1982). Its function remains to be determined, and since it lacks phosphotyrosine, it is unlikely to be a substrate of the tyrosyl kinase activity of pp60^{src}.

pp50 is not as highly conserved in structure in avian and mammalian cells or as abundant as pp90 (OPPERMANN et al. 1981a). A phosphoserine-containing form of pp50 can be detected in normal or *td*RSV-infected CEF cells, whereas a phosphoserine- and phosphotyrosine-containing form identical to the complexed form of pp50 can be detected in *nd*RSV-infected cells (BRUGGE and DARROW 1982; GILMORE et al. 1982). Although the tyrosine phosphorylation of pp50 is transformation dependent, it is not temperature sensitive in cells infected by *ts*NY68 or *ts*LA29. This behavior is different from that of other putative substrates of pp60^{src}. One explanation for this anomaly is that pp50 is a high-affinity substrate, whose phosphorylation is unimpaired in cells infected by temperature-sensitive mutants in which the pp50:pp60^{src}:pp90 complex is extremely stable (BRUGGE et al. 1981;

BRUGGE and DARROW 1982). However, the tyrosine phosphorylation of pp50 may have no biological significance: pp50 has no phosphotyrosine in RSV-transformed mammalian cells or in CEF cells infected by mutant *ts*CH119 (BRYANT and PARSONS 1982; J. BRUGGE, personal communication). Antibodies specific to pp50 are not yet available to probe the precise nature of its interaction with pp60^{src}, although it appears to form a complex with pp90 even in uninfected cells (BRUGGE et al. 1983). However, antiserum specific for the carboxy-terminus of pp60^{src} immunoprecipitates the complex less efficiently than TBR serum (SEFTON and WALTER 1982), suggesting either that pp50 and pp90 bind to the carboxy-terminus of pp60^{src} or that binding of these cellular proteins to pp60^{src} alters the conformation of the carboxy-terminal domain of pp60^{src}. The apparent defect in immune complex kinase activity of bound pp60^{src} may be caused by such an alteration. pp50 and pp90 also form complexes with the transforming proteins of FSV and Y73 sarcoma viruses (LIPSICH et al. 1982), suggesting a common role for the interaction of these cellular proteins with tyrosyl kinase-encoding viral transforming proteins.

Pulse-chase and fractionation studies (COURTNEIDGE and BISHOP 1982; BRUGGE et al. 1983) suggested that newly synthesized pp60^{src} is found to preferentially associate with pp50 and pp90 in a short-lived complex (half-life of about 7–15 min, depending on the strain of RSV), and that the complex is found in the cytosol fraction, while the free form of pp60^{src} is found in the particulate fraction of RSV-infected cells. The kinetics of dissociation of the complex are altered in *ts*NY68-infected cells: the complex is very stable, and little free, unbound pp60^{src} is present in these cells at the restrictive temperature. Most of the pp60^{src} in *ts*NY68-infected cells at the restrictive temperature is in the soluble fraction, while the small amount of membrane-associated pp60^{src} observed upon fractionation appears still to be associated with the complex. Temperature shift-down from 41° to 35° in the presence of cycloheximide of pulse-labeled *ts*NY68-infected cells followed by cell fractionation suggests that pp60^{src} in the cytoplasmic complex is the precursor to membrane-associated monomer pp60^{src}; and the converse experiment suggests that at least a portion of membrane-associated pp60^{src}can reenter the soluble complex upon temperature shift-up (COURTNEIDGE and BISHOP 1982). These observations have led to a model for the biogenesis of pp60^{src}, illustrated in Fig. 13, in which the pp50:pp60^{src}:pp90 complex is the vehicle by which pp60^{src} is transported through the cytoplasm to the plasma membrane. The exact nature of the association between pp60^{src} and the preexisting pp50:pp90 complex is not clear. Different strains of RSV have different kinetics of dissociation (BRUGGE et al. 1981, 1983). Although many temperature-sensitive mutants have essentially all of their pp60^{src} in the soluble complex at the restrictive temperature (COURTNEIDGE and BISHOP 1982; BRUGGE et al. 1983), this is not the case for *ts*CH119-infected CEF cells, in which pp60^{src} is chiefly membrane associated at either temperature (BRYANT and PARSONS 1982). The interpretation of experiments performed with temperature-sensitive mutants requires determination of whether the decreased membrane association

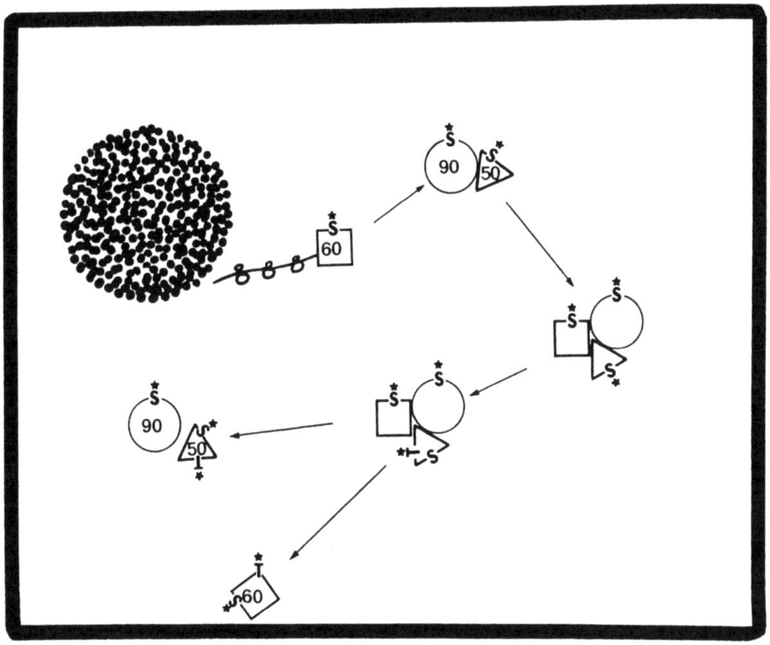

Fig. 13. Schematic model for biosynthesis of pp60^{src} on free polysomes and its transport to the plasma membrane. Shortly after synthesis, pp60^{src} is phosphorylated on serine and associates with a preexisting pp50:pp90 complex, both phosphorylated on serine (S^*). After pp60^{src} binds to pp50:pp90, it forms a stable but transient complex and pp50 is phosphorylated on tyrosine (T^*). pp60^{src} is transported to the plasma membrane bound to the pp50:pp90 complex. After pp60^{src} reaches the membrane, it integrates into the lipid bilayer and the pp50:pp90 complex is released. It is not known at which point lipid is bound to pp60^{src}. [The model shown was suggested by J. BRUGGE; a similar model has been suggested by COURTNEIDGE et al. (1982)]

of pp60^{src} is due to an intrinsic property of the mutant pp60^{src} or whether it is due to stable binding to pp50 and pp90. Association of lipid with pp60^{src} also appears to be involved in membrane association (see Sect. 5), but it is not yet clear when this event occurs in the postulated chronological sequence in the transport of pp60^{src} to the plasma membrane. The interaction of pp60^{src} encoded by mutant and variant RSVs having altered membrane association (see Sect. 4) with the pp50:pp90 complex will have to be examined to determine the accuracy of the model. The precise physiological role of the complex in modulating pp60^{src} function has yet to be established.

8 Putative Substrates of pp60^{src}

8.1 pp60^{src} Is a Tyrosyl Kinase: Strategies for Identifying Substrates

The possession by pp60^{src} of intrinsic protein kinase activity is suggested by the following lines of evidence: (a) kinase activity copurifies with pp60^{src}

(ERIKSON et al. 1979; MANESS et al. 1979; LEVINSON et al. 1980; (b) protein kinase activity is thermolabile in mutants temperature sensitive in the *src* gene (COLLETT and ERIKSON 1978; LEVINSON et al. 1978; RUBSAMEN et al. 1979; SEFTON et al. 1980a); (c) the cell-free translation product of the *src* gene possesses protein kinase activity (ERIKSON et al. 1978; SEFTON et al. 1979); and (d) pp60^{src} produced in *Escherichia coli* from recombinant molecular clones has protein kinase activity (GILMER and ERIKSON 1981; McGRATH and LEVINSON 1982). The kinase activity of pp60^{src} is specific for tyrosine residues on substrate proteins (HUNTER and SEFTON 1980; COLLETT et al. 1980; LEVINSON et al. 1980). Not only is this an unusual amino acid substrate specificity, but phosphotyrosine is an extremely rare protein modification in normal cells. Transformation by RSV results in a tenfold elevation in the total level of phosphotyrosine in cellular protein (HUNTER and SEFTON 1980; SEFTON et al. 1980b). This elevation is largely due to phosphorylation of cellular proteins by pp60^{src}, and is temperature sensitive in RSV-transformed cells containing a temperature-sensitive mutation in the *src* gene (SEFTON et al. 1980b). Transformation by RSV might occur via pp60^{src}-mediated phosphorylation on tyrosine residues of a variety of cellular proteins, thereby explaining the pleiomorphism of the transformed phenotype. It has been suggested that the criterion of tyrosine phosphorylation of a cellular protein modulated in transformation-dependent manner may be used to identify substrates of pp60^{src} (SEFTON et al. 1980b; HUNTER 1980; HUNTER et al. 1981). Four basic approaches have been taken to identify potential pp60^{src} substrates. These approaches will be discussed in Sects. 8.1.1–8.1.4, while individual putative substrates will be discussed in Sects. 8.2–8.5.

A central issue in any of the approaches to identifying substrates of pp60^{src} is whether phosphotyrosine-containing proteins in RSV-transformed cells are phosphorylated directly by pp60^{src} or whether the phosphorylations are mediated by cellular tyrosyl protein kinases which are induced or modified by pp60^{src}. A close kinetic correlation between increased tyrosine phosphorylation and reactivation of pp60^{src} kinase activity early in the process of transformation upon temperature shift-down of cells infected by RSV mutants temperature sensitive in the *src* gene may provide a good argument that a protein is a direct target of pp60^{src} and not a target of a cellular tyrosyl kinase induced by pp60^{src}, but such demonstrations may be technically difficult to achieve. Demonstration of the ability of purified pp60^{src} to phosphorylate a purified protein identified as a substrate may be another argument for substrate authenticity, but this approach is subject to the criticism of the questionable nature of a kinase's in vitro substrate specificity. Colocalization of a potential substrate with pp60^{src} is suggestive, but not persuasive, evidence, since the kinase-active domain of membrane-associated pp60^{src} (see Fig. 11) could act on membrane-associated or soluble substrates. Correlation between tumorigenicity and tyrosine phosphorylation of a putative substrate in cells transformed by mutant viruses may provide evidence that the phosphorylation has physiological significance relative to transformation, but this may also be difficult to demonstrate.

8.1.1 Coimmunoprecipitation with pp60^{src}

One method that has been used to identify substrates of pp60^{src} is to examine cellular phosphoproteins that coimmunoprecipitate with pp60^{src} due to their physical association with it. This approach has identified two proteins, pp90 and pp50, that form a complex with pp60^{src} (Hunter and Sefton 1980; Brugge et al. 1981; Oppermann et al. 1981a). The complex appears to be involved in transport of pp60^{src} to the plasma membrane (Brugge et al. 1983; Courtneidge and Bishop 1982). These proteins were discussed in greater detail in Sect. 7. For a cellular protein that associates with pp60^{src} to be detected by the method of coimmunoprecipitation, the physical interaction must be stable and relatively long-lived and the complex must be relatively abundant. Transient associations of pp60^{src} with low abundance phosphoproteins may be difficult to detect. The conclusion that coimmunoprecipitating cellular proteins are physically associated with pp60^{src} must be supported by demonstrations of cosedimentation or cofractionation, lack of structural relationship to pp60^{src}, and demonstration that the complex is not an artifact of cell lysis. Demonstration of the physical reality of complex formation with pp60^{src} does not guarantee physiological significance.

8.1.2 Comparison of Phosphoproteins Resolved by Electrophoresis

Another approach involves comparison by one- or two-dimensional gel electrophoresis of phosphoproteins from normal and RSV-transformed cells or from cells infected with RSV conditional mutants grown at the permissive and restrictive temperatures. Phosphotyrosine content of transformation-specific candidate substrates can be verified by phosphoamino acid analysis, but the detection of phosphoproteins containing phosphotyrosine is aided by a procedure based upon the greater alkali resistance of the phosphoester bond of phosphotyrosine (Cooper and Hunter 1981a; Hunter et al. 1981; Cheng and Chen 1981a, b). Treatment of ^{32}P-labeled phosphoproteins resolved in polyacrylamide gels with alkali hydrolyzes the phosphoester bonds of the majority of the phosphoserine residues, and since phosphoserine is the most abundant phosphoamino acid in cellular protein (Hunter and Sefton 1980), its removal greatly enhances the probability of detecting the rare phosphotyrosine-containing protein. Although this technique is useful, it may not detect all proteins containing phosphotyrosine: (a) some proteins contain alkali-sensitive tyrosine-phosphate linkages, (b) low-abundance phosphoproteins may not be detected because of labeling considerations or rapid turnover at a tyrosine phosphorylation site, and (c) proteins containing alkali-resistant phosphoserine residues might obscure detection of a similarly migrating tyrosine-phosphorylated protein (Cooper and Hunter 1981a; Martinez et al. 1982).

One-dimensional gel electrophoresis of a lysate of ^{32}P-labeled cells extracted directly into boiling SDS followed by phosphoamino acid analysis of gel slices suggests that there are at least 30 phosphotyrosine-containing proteins in RSV-transformed CEFs, of which a 59-kd (probably pp60^{src} itself) and a 36-kd protein are the most abundant (MARTINEZ et al. 1982). A similar one-dimensional analysis of a radioimmune precipitation buffer (RIPA)-solubilized cell extract established a minimum of ten potential substrates (BEEMON et al. 1982). No difference in the pattern of phosphoproteins present in normal and RSV-transformed cells was discernible on this analysis. This type of analysis also has been used to examine tyrosine phosphorylation of specific proteins after mitogen stimulation of normal CEF cells by epidermal growth factor, platelet-derived growth factor, and multiplication-stimulating activity (NAKAMURA et al. 1983). This procedure is limited by the resolving power of one-dimensional gels; and it provides no information beyond phosphotyrosine content of a given protein.

Separation of total cellular phosphoproteins by two-dimensional gel electrophoresis has identified several potential pp60^{src} substrates: a 36-kd protein (RADKE and MARTIN 1979, 1980; RADKE et al. 1980; KOBAYASHI and KAJI 1980), and 46-kd and 28-kd proteins (COOPER and HUNTER 1981a, b) whose tyrosine phosphorylation is increased by RSV transformation. These proteins will be discussed in Sect. 8.2 and 8.3. Two acidic membrane phosphoproteins (57 kd and 55 kd) specifically dephosphorylated upon RSV transformation of CEF cells have been identified by two-dimensional gel electrophoresis (WITT and GORDON 1980), but since pp60^{src} contains no identified phosphatase activity, these proteins will not be considered herein as pp60^{src} substrates. Although two-dimensional gel electrophoresis has high resolving power and sensitivity, the technique has several limitations: (a) the loading capacity of isoelectric focusing gels is low, so that low-abundance phosphoproteins may not be detected; (b) some phosphoproteins (such as pp60^{src}) do not electrofocus sharply; (c) artifacts may be introduced during sample and gel preparation (see O'FARRELL 1975; O'FARRELL et al. 1977); (d) unless a protein's phosphate labeling content is altered markedly upon transformation, or changes in relative radiolabeling intensity are monitored carefully, potential substrates may not be detected; and (e) no information other than molecular weight and isoelectric point is provided.

8.1.3 Proteins of Known Function

Since the coimmunoprecipitation and gel comparison techniques do little to illuminate the functions of the potential substrates identified, another approach is to analyze proteins whose function is known and is affected (directly or indirectly) by RSV transformation, particularly elements of the cytoskeleton (see HANAFUSA 1977; BAUER et al. 1982). The ability of purified pp60^{src} to phosphorylate in solution exogenously added substrates has been tested (ERIKSON et al. 1979, 1980; MANESS et al. 1979; LEVINSON et al. 1980; GILMER and ERIKSON 1981; ITO et al. 1982; RICHERT et al. 1982): TBR IgG heavy chain, casein, tubulin (alpha and beta subunits), actin, vimentin, alpha-actinin, vinculin, and the 10-nm filament protein desmin could serve as substrates, while bovine serum albumin, ovalbumin, cytochrome *c*, protamine, histones (lysine rich and arginine rich), phosvitin, myosin, tropomyosin, pyruvate kinase, and phosphofructokinase could not. Purified pp60^{src} also can phosphorylate the nonprotein substrate glycerol (RICHERT et al. 1982; GRAZIANI et al. 1983), but not glucose, fructose, adenosine, or deoxyadenosine (RICHERT et al. 1982). Since phosphorylation of exogenous substrates in vitro may not be representative of true substrate specificity in vivo (KREBS and BEAVO 1979), the proteins phosphorylated in vitro by purified pp60^{src} need not be authentic substrates. To search for authentic substrates (defined operationally as those proteins whose phosphotyrosine content increases upon transformation), proteins from normal or RSV-transformed cells labeled biosynthetically with [^{32}P]orthophosphate are immunoprecipitated with monospecific antisera, purified by SDS-polyacrylamide gel electrophoresis, and analyzed for phosphoamino acid content. Using this approach, vinculin has been identified as a cytoskeletal target of pp60^{src}, while the cytoskeletal proteins filamin, myosin heavy chain, alpha-actinin, vimentin, fibronectin, tubulin, and actin appear not to be pp60^{src} substrates (SEFTON et al. 1981; HUNTER et al. 1981; SEFTON et al. 1982a). Vinculin phosphorylation will be discussed in Sect. 8.4. This approach to identifying substrates is limited to proteins to which specific antisera are available.

8.1.4 In Vitro Systems

Yet another approach to identifying substrates involves in vitro systems in which isolated components from normal or RSV-transformed cells are incubated with gamma-labeled [^{32}P]ATP, proteins radiolabeled in situ analyzed, and transformation-specific phosphotyrosine-containing proteins identified. These systems allow the investigator to analyze the topological relationships between the pp60^{src} protein kinase in situ and endogenous phosphate acceptor proteins in close proximity to it in cytoskeletal framework preparations (BURR et al. 1980; 1981), isolated adhesion plaques (SHRIVER and ROHRSCHNEIDER 1981b; ROHRSCHNEIDER et al. 1982), or membrane fractions (GOLDBERG et al. 1980; GALLIS et al. 1981; GARBER et al. 1983a). A number of potential substrates have been identified by these systems, and will be discussed in Sect. 8.5. A basic rationale for developing in vitro systems is that dissection of the molecular mechanisms of cellular transformation mediated by pp60^{src} will be difficult in the intact cell. There are several limitations to such systems: (a) Implicit in their development is the assumption that pp60^{src} localization and function are strongly related and that authentic substrates will colocalize with pp60^{src} in membranes or cytoskeletal structures. However, it is entirely possible that substrates may be soluble proteins. (b) Phosphorylatable sites in substrates may already be occupied by unlabeled phosphate and not be readily exchangeable. (c) Steric inaccessibility, absence of required factors, or presence of inhibitory factors may impose constraints or introduce artifacts. (d) Substrate specificity in vitro may not reflect that of pp60^{src} in the intact cell.

8.2 The 36-kd Transformation-Specific Phosphoprotein

Using two-dimensional polyacrylamide gel electrophoresis of radiolabeled cell lysates to examine the effects of RSV transformation on the pattern of synthesis and phosphorylation of cellular proteins, RADKE and MARTIN (1979) detected a 36-kd protein whose phosphorylation is an early event in cellular transformation of CEF cells. The 36-kd phosphoprotein was detectable within 20 min after a shift from the nonpermissive to permissive temperature of CEF cells infected by the RSV mutant *ts*NY68. Protein synthesis was not required for the appearance of the 36-kd phosphoprotein in temperature-shift experiments (RADKE and MARTIN 1980). Phosphorylation of the 36-kd protein also was an early event in the activation of the *src* gene product upon temperature shift-down of chick myoblasts infected with *ts*NY68 (KOBAYASHI and KAJI 1980). An unphosphorylated precursor 36-kd protein has been indentified in normal cells, and the phosphorylated 36-kd protein in transformed cells (avian and mammalian) contains phosphotyrosine (RADKE et al. 1980; ERIKSON and ERIKSON 1980); but it is estimated that only 5%–10% of the 36-kd protein is phosphorylated in transformed cells. The 36-kd protein can be phosphorylated on tyrosine in vitro by a kinase attributable to $pp60^{src}$ in crude cell extracts (KOBAYASHI et al. 1981) or in purified plasma membranes from RSV-transformed cells (AMINI and KAJI 1983). The purified 36-kd protein can be phosphorylated in vitro by purified $pp60^{src}$ at the same tyrosine residue as the major site of tyrosine phosphorylation in vivo (ERIKSON and ERIKSON 1980), suggesting that it is a substrate for the protein kinase activity of $pp60^{src}$. One-dimensional gel analysis revealed a peak of phosphotyrosine at a molecular weight region corresponding to the 36-kd phosphoprotein (MARTINEZ et al. 1982; BEEMON et al. 1982). This peak of phosphotyrosine was absent from the phosphotyrosine profile derived from cells infected by the RSV partial transformation mutant CU2, which is defective in 36-kd protein phosphorylation (MARTINEZ et al. 1982; NAKAMURA and WEBER 1982). The 36-kd phosphoprotein has also been detected as an alkali-resistant phosphotyrosine-containing protein (COOPER and HUNTER 1981a; HUNTER et al. 1981) that is also present in avian and mammalian cells transformed by other unrelated retroviruses whose transforming gene products possess tyrosyl protein kinase activity (ERIKSON et al. 1981a; COOPER and HUNTER 1981b), suggesting that it is a highly conserved protein whose phosphorylation is associated with the mechanism of transformation by certain retroviruses. The same 36-kd protein is also phosphorylated on tyrosine in A-431 human epidermoid carcinoma cells in response to epidermal growth factor treatment (HUNTER and COOPER 1981; COOPER and HUNTER 1981c; ERIKSON et al. 1981b), suggesting a common link in the series of biochemical events caused by RSV transformation and EGF stimulation, although the exact congruence of the tyrosine phosphorylation sites (although they are on the same tryptic peptide) has not been established.

No function has been assigned to the 36-kd protein. It has no carbohydrate component, and it is not secreted by transformed cells (ERIKSON and

ERIKSON 1980; COURTNEIDGE et al. 1983). There has been a report that partially purified 36-kd phosphoprotein was associated with cytosolic malate dehydrogenase activity (RUBSAMEN et al. 1982), but this could not be confirmed (COURTNEIDGE et al. 1983; COOPER et al. 1983b; GREENBERG and EDELMAN 1983a), and several lines of evidence suggest that the 36-kd protein is not malate dehydrogenase: (a) antibody to glycolytic enzymes and antibody to purified 36-kd proteins immunoprecipitate proteins with different electrophoretic mobilities; (2) the 36-kd protein does not bind to Cibachron blue or AMP-agarose under conditions in which many dehydrogenases bind; (3) malate dehydrogenase is an acidic protein, whereas the 36-kd protein is basic; and (d) preliminary sequencing data suggests that the proteins are different (COOPER and HUNTER 1982b).

Using two-dimensional gel electrophoresis to assess the presence or absence of phosphorylated 36-kd protein, KRUEGER et al. (1982) observed no obvious correlation between 36-kd protein phosphorylation and tumorigenicity in wild-type RSV and several rASVs. The 36-kd phosphoprotein can also be detected in RR1022 cells and in CEF cells transformed by the rescued virus RSV-RR (GARBER, unpublished work). However, using the more sensitive assay of immunoprecipitation with antiserum to purified 36-kd protein, WEBER and co-workers (NAKAMURA and WEBER 1982; KAHN et al. 1982) examined the degree of correlation between 36-kd protein phosphorylation and tumorigenicity of cells infected with partial transformation mutants of RSV. The observations of these workers suggest that phosphorylation of the 36-kd protein is: (a) neither necessary nor sufficient for loss of surface fibronectin and loss of density-dependent inhibition of growth; (b) insufficient for morphological alterations, loss of adhesiveness, and increase in 2-deoxyglucose transport; (c) perhaps necessary, but not sufficient for tumorigenicity; and (d) perhaps sufficient for an increase in plasminogen activator.

The 36-kd phosphoprotein was detected in cytoskeletal frameworks prepared from RSV-transformed cells by extraction in situ with nonionic detergent-containing buffers (CHENG and CHEN 1981a, b; COOPER and HUNTER 1982a, b; GREENBERG and EDELMAN 1983b), but was found in a soluble cellular fraction when transformed cells in suspension were extracted with nonionic detergents (COOPER and HUNTER 1982a, b). Biochemical fractionation indicates that the 36-kd protein is located in plasma membrane fractions (COURTNEIDGE et al. 1983; AMINI and KAJI 1983; GREENBERG and EDELMAN 1983a, b), but the degree of association of the protein with the particulate fraction in transformed cell homogenates was found to be dependent upon salt concentration and the presence of divalent cations (COOPER and HUNTER 1982a; COURTNEIDGE et al. 1983; GREENBERG and EDELMAN 1983b), suggesting that it is a peripheral membrane protein, perhaps only loosely associated with the plasma membrane. In vitro translation and pulse-chase studies suggest that the 36-kd protein is synthesized in the cytoplasm as a 36-kd primary translation product that is rapidly transported to the membrane, and is not an integral membrane protein (COOPER and HUNTER 1982b; COURTNEIDGE et al. 1983). Immunofluorescent studies with specific antiserum suggest that the 36-kd protein is attached to the cytoplasmic aspect of the plasma membrane, but that its intracellular distribution is dissimilar from cytoskeletal patterns (COOPER and HUNTER 1982a, b; COURTNEIDGE et al. 1983; GREENBERG and EDELMAN 1983b). The staining pattern is similar to that of the membrane-associated protein alpha spectrin (GREENBERG and EDELMAN 1983b). The localization of the 36-kd protein as a peripheral membrane protein is not inconsistent with its being a substrate of $pp60^{src}$, an integral membrane protein, but the lack of strong correlation between 36-kd protein phosphorylation and membrane-associated transformation parameters, especially tumorigenicity (NAKAMURA and WEBER 1982; KAHN et al. 1982), casts doubts upon its physiological significance.

8.3 The 46-kd and 28-kd Alkali-Resistant Putative Substrates

Using two-dimensional electrophoresis of ^{32}P-labeled CEF cell lysates and alkali treatment of gels, COOPER and HUNTER (1981a) and HUNTER et al. (1981) detected seven alkali-resistant phosphotyrosine-containing proteins in which the level of phosphorylation was increased upon RSV transformation. Three of these (46 kd:pI 7.05, 39 kd:pI 7.31, and 39 kd:pI 7.33) were not readily detectable in untransformed cells, whereas the other four (46 kd:pI 6.95, 43 kd:pI 6.80, 43 kd:pI 7.32, and 28 kd:pI 7.38) underwent a marked elevation in phosphorylation. All of these putative substrates contained phosphoserine as well as phosphotyrosine. The most prominent protein detected by this method was the 36-kd protein described in Sect. 8.2 (identified as 39 kd:pI 7.33 by COOPER and HUNTER). Increased phosphorylation of the seven candidate substrate proteins did not occur at the restrictive temperature in cells infected by the RSV temperature-sensitive mutant *ts*LA29, but no detailed temperature-shift experiments were described to establish whether phosphorylations other than that of the 36-kd protein were early events in transformation, a demonstration that would have strengthened the argument that these proteins are true $pp60^{src}$ substrates. Precursor-product relationships between the phosphorylated putative substrates and non-phosphotyrosine-containing normal cell proteins could be tentatively established only for the 39 kd:pI 7.33, 46 kd:pI 7.05, and 28 kd:pI 7.38 proteins (COOPER and HUNTER 1981a, 1983). Furthermore, only a small percentage of the precursor proteins become phosphorylated in transformed cells. The same seven transformation-specific phosphoproteins were detected in chicken cells transformed by other retroviruses (FSV, PRCII, or Y73); the 39 kd:pI 7.33 and 28 kd:pI 7.38 phosphoproteins were found in both mammalian and avian transformed cells; but only the phosphorylations of the 46 kd:pI 7.05, 43 kd:pI 7.32, and 39 kd:pI 7.33 proteins were temperature sensitive in cells infected by a temperature-sensitive FSV mutant (COOPER and HUNTER 1981b). Attention has been focused on the 46 kd:pI 7.05, 39 kd:pI 7.33, and 28 kd:pI 7.38 phosphoproteins, as these are not readily detected in uninfected cells (although candidate nonphosphorylated forms have been identified), but contain a high proportion of phosphotyrosine relative to phosphoserine and phosphothreonine in transformed cells (COOPER and HUNTER 1982a, b, 1983).

The 46-kd and 28-kd phosphoproteins and their presumptive precursors were present predominantly in the soluble high-speed supernatant fraction in transformed CEF cells homogenized in a Teflon tissue grinder in hypotonic phosphate-containing buffer (COOPER and HUNTER 1982a, b). The distribution of the two proteins was unaffected by the presence of the divalent cation chelator EDTA. The distribution of enzyme markers or known membrane-associated proteins was not monitored in these experiments (see comments in Sect. 11.1.3). The 46-kd and 28-kd proteins were also extracted into the soluble fraction upon Triton X-100 extraction of transformed cells in situ or in suspension to obtain cytoskeletal preparations. The 46-kd protein is also soluble in RSV-transformed mammalian cells (COOPER and HUNTER 1982b). Preliminary in vitro translation and pulse-chase studies suggest that the 46-kd protein is not synthesized as a higher molecular weight precursor; and preliminary immunofluorescent studies confirm the cytoplasmic distribution of the 46-kd protein (COOPER and HUNTER 1982a, b). If the 46-kd and 28-kd proteins are true pp60^{src} substrates, then there is no obvious pattern to the localization of tyrosyl protein kinases and their substrates.

Use of antisera to purified chicken breast muscle glycolytic enzymes has enabled the 46-kd and 28-kd proteins to be identified as the glycolytic enzymes enolase and phosphoglycerate mutase respectively (COOPER et al. 1983b). This suggests that tyrosine phosphorylation of glycolytic enzymes could contribute to the high rate of aerobic glycolysis in RSV-transformed cells. Phosphorylation of the 46-kd and 28-kd proteins and transformation-associated parameters have been examined in cells transformed by partial transformation mutants of RSV (COOPER et al. 1983a). Results suggest that phosphorylation of the 46-kd protein is (a) neither necessary nor sufficient for changes in adhesiveness and morphology. (b) unnecessary for increases in 2-deoxyglucose transport and plasminogen activator, and (c) perhaps necessary, but not sufficient for anchorage independence. Phosphorylation of the 28-kd protein is (a) neither necessary nor sufficient for changes in adhesiveness and morphology, (b) insufficient for increase in plasminogen activator, (c) perhaps sufficient for increase in 2-deoxyglucose transport, and (d) perhaps necessary, but not sufficient for anchorage independence. However, phosphorylation of the 46-kd and 28-kd proteins could be adventitious, and the evidence that these proteins are true substrates of pp60^{src} is less convincing than that for the 36-kd phosphoprotein.

8.4 Vinculin

Since the organization of the cytoskeleton is affected profoundly by RSV transformation, an examination of the proteins of the cytoskeleton is an obvious starting point in a search for pp60^{src} substrates. To test whether the level of phosphotyrosine in cytoskeletal proteins was increased in transformed cells, proteins from normal or RSV-transformed CEF cells labeled for long periods with [^{32}P]orthophosphate (to ensure steady-state labeling of phosphoamino acids) were immunoprecipitated with specific antisera, purified by SDS-PAGE, and analyzed for phosphoamino acid content (SEFTON et al. 1981; HUNTER et al. 1981; SEFTON et al. 1982a). Actin, tubulin, alpha-actinin, fibronectin, filamin, myosin heavy chain, vimentin, and vinculin were examined, but only vinculin showed a marked (tenfold) transformation-dependent increase in phosphotyrosine content. In cells infected with the temperature-sensitive transformation mutant *ts*LA29, the level of tyrosine phosphorylation in vinculin was high at the permissive temperature and low at the nonpermissive temperature. The report did not include data on the kinetics of increased phosphorylation of tyrosine on vinculin after shift-down to the permissive temperature. Vinculin can be phosphorylated in vitro by partially purified pp60^{src} (ITO et al. 1982). The data indicate that vinculin may be a good candidate for a substrate of pp60^{src}. Elevation of phosphotyrosine in vinculin was also observed in avian cells transformed by Y73 virus and mammalian cells transformed by RSV or Ab-MuLV, but no elevation was observed in avian cells transformed by chemicals or by PRCII virus or in mammalian cells transformed by SV40. Vinculin tyrosine phosphorylation is also elevated in avian cells transformed by the *fps*-transforming gene-encoding viruses UR1 and 16L, but not in cells transformed by RSV-RR, FSV, or UR2 viruses (ANTLER et al. to be published). These observations on one hand suggest that transformation per se does not lead to activation of a cellular tyrosyl kinase that phosphorylates vinculin, but on the other hand cast doubts upon the validity of a unified theory of substrate selection by viral tyrosyl kinases.

Vinculin is a 130-kd protein that is concentrated at sites of actin-membrane interaction and cell-substratum contact. In normal cells, it is enriched at the ends of actin-containing microfilament bundles, where the bundles make contact with the membrane at focal adhesion plaques (see review by HYNES 1982). In RSV-transformed cells, focal contacts are fewer and actin, alpha-actinin, and vinculin have altered distributions. It has been suggested that the increased tyrosine phosphorylation of vinculin upon RSV transformation modulates its function, altering the organization and membrane association of microfilaments and the interaction with extracellular fibrils of fibronectin, leading to the phenotypic changes characteristic of the transformed state (HUNTER et al. 1981). Vinculin colocalizes with pp60^{src} in SR-NRK cells and can be phosphorylated in vitro in isolated adhesion plaques (SHRIVER and ROHRSCHNEIDER 1981a, 1981b). The localization of vinculin within the adhesion plaques was not temperature sensitive, however, in NRK cells infected by RSV mutant *ts*LA91 (SHRIVER and ROHRSCHNEIDER 1981b). Vinculin also colocalizes with pp60^{src} within the adhesion plaques in RSV B77-transformed NRK cells (SEFTON et al. 1982a), again supporting its identification as a potential substrate. The demonstration that the in vivo phosphorylated electrophoretic isoform of vinculin is predominantly found in the Triton-insoluble fraction (membrane- and cytoskeleton-bound vinculin) in chicken gizzard cells (GEIGER 1982) lends further support to the notion that only a small fraction of the total vinculin molecules can have a major functional role. The phosphotyrosine content in vinculin and the localization of vinculin within adhesion plaques have been studied in CEF cells infected with partial transformation mutants of RSV (ROHRSCHNEIDER et al. 1982; ROHRSCHNEIDER and ROSOK 1983; see also Sect. 6). Vinculin was found within adhesion plaques regardless of the virus used to infect cells and regardless of the temperature of growth. Vinculin tyrosine phosphorylation was temperature sensitive in cells infected by *ts*NY68 or *ts*CU11, but there was no elevation in tyrosine phosphorylation of vinculin in cells infected by mutant CU2, despite the overabundance of both vinculin and $pp60^{src}$ in adhesion plaques in CU2-infected cells. The same correlations drawn between 36-kd protein phosphorylation and transformation-associated parameters that were discussed in Sect. 8.2 would seem to apply to correlations between vinculin phosphorylation and transformation-associated parameters, suggesting that the data are not sufficient to make a strong conclusion that $pp60^{src}$-mediated tyrosine phosphorylation of vinculin has great physiological significance. Furthermore, increased tyrosine phosphorylation of vinculin does not occur during the release of stress fibers before mitosis in normal cells (ROSOK and ROHRSCHNEIDER 1983), suggesting that vinculin phosphorylation is unrelated to the maintenance of intact stress fibers.

8.5 Proteins Phosphorylated In Vitro Systems

If the topographical relationships between $pp60^{src}$ and potential substrates are preserved, transformation-specific phosphotyrosine-containing proteins can be radiolabeled in vitro when isolated components from RSV-transformed cells are incubated with gamma-labeled [^{32}P]ATP. BURR and coworkers (1980, 1981) have described a protein kinase activity attributable to $pp60^{src}$ that is associated with the Triton-insoluble fraction of RSV-infected CEF cells. This activity phosphorylates $pp60^{src}$ and proteins of 50 kd, 68 kd, and 95 kd on tyrosine in cytoskeletal frameworks. Upon gentle extraction with Triton-containing buffer, virtually all of the $pp60^{src}$ in the cell remains associated with the Triton-insoluble fraction, while the cytoplasmic glycolytic enzymes pyruvate kinase, lactate dehydrogenase, and hexokinase are extracted into the Triton-soluble fraction. When the Triton-insoluble fraction is incubated with gamma-labeled [^{32}P]ATP in an in situ phosphorylation reaction, phosphorylation of proteins of 60 kd and 68 kd are particularly prominent in preparations from RSV-transformed cells, but absent or greatly reduced in preparations from uninfected or *td*RSV-infected cells. The 60-kd protein comigrates with authentic $pp60^{src}$, is immunoprecipitated by TBR serum, is phosphorylated on a tyrosine residue in the *S. aureus* V8 protease carboxy-terminal 26-kd peptide, and this phosphorylation is temperature sensitive in cells infected by *ts*NY68. The phosphorylation of $pp60^{src}$ in situ is inhibited by Ca^{+2}, as was $pp60^{src}$ phosphorylation in detergent-disrupted cell extracts (LEVINSON et al. 1978). The function of the 68-kd protein is unknown, but its tyrosine phosphorylation is an early event in transformation, assayed in cytoskeletons from *ts*NY68-infected cells following temperature shift-down. The 68-kd protein may be fimbrin, a 68-kd cytoskeletal protein shown to colocalize with pp60^{src} within adhesion plaques in SR-NRK cells (SHRIVER and ROHRSCHNEIDER 1981b). Another prominently labeled protein of 95 kd, which was phosphorylated to about the same extent in both normal and transformed cells, was found to contain phosphotyrosine. The function of the 95-kd protein is as yet unknown. A protein of 50 kd which coimmunoprecipitates with pp60^{src} was also labeled in RSV-transformed CEF Triton-insoluble fractions. It has not yet been determined whether this protein is identical to the 50-kd protein present in a complex with pp60^{src} (see Sect. 7). The 50-kd, 68-kd, and 95-kd proteins labeled in cytoskeletal fractions are candidate substrates, but their role in transformations has yet to be established.

SHRIVER and ROHRSCHNEIDER (1981b) have described a tyrosyl kinase activity attributable to pp60^{src} that is associated with isolated adhesion plaques and junctions from SR-NRK cells. This activity phosphorylates a number of proteins. Immunofluorescence experiments revealed that adhesion structures isolated by gentle detergent extraction and shear force applied to monolayer cultures of SR-NRK cells contained pp60^{src}, actin, alpha-actinin, vinculin, and fimbrin distributed in a fashion similar to that observed at the ventral surface of fixed whole cells, but lacked tubulin and the 58-kd intermediate filament protein. Two-dimensional gel electrophoretic analysis of the products radiolabeled in an in situ phosphorylation reaction, followed by alkali treatment of the gel and phosphoamino acid analysis of radiolabeled spots, identified at least ten alkali-resistant phosphoproteins, of which at least seven contained phosphotyrosine (150 kd, 145 kd, 130 kd, 73 kd, 70 kd, a tetrad of 60-kd species, and 48 kd). The 60-kd species were shown to be pp60^{src}: they were immunoprecipitated by TBR and contained phosphotyrosine on the carboxy-terminal 26-kd *S. aureus* V8 protease peptide. The 130-kd protein was shown to be vinculin: partial proteolytic peptides from the 130-kd spot and authentic vinculin immunoprecipitated with specific antiserum were identical, and vinculin immunoprecipitated with specific antiserum from the reaction products or from metabolically labeled cells comigrated with the 130-kd spot. The identity and functions of the other phosphotyrosine-containing candidate substrates are as yet unknown. The kinase activity in adhesion plaques from *ts*LA91-transformed NRK cells is thermolabile. pp60$^{c\text{-}src}$ was also phosphorylated in adhesion structures from both normal and RSV-transformed NRK cells (ROHRSCHNEIDER et al. 1982). Colocalization of phosphorylatable pp60^{src} and vinculin highlights adhesion plaques as an important site of pp60^{src} action.

GALLIS and co-workers (1981) have described a tyrosyl kinase activity in membrane vesicles isolated from RSV-transformed rat cells (RS-1) that phosphorylates proteins of 37 kd, 50 kd, and 67 kd. The RS-1 line of SR-RSV-D-transformed rat cells used in this study has not been characterized with respect to pp60^{src} localization. Although membrane vesicles from RS-1 contain pp60^{src} immune complex kinase activity, it has not been demonstrated that $pp60^{src}$ is chiefly membrane associated in these cells. In vitro phosphorylation reactions with RS-1 membrane vesicles resulted in the radiolabeling of proteins of 37 kd, 50 kd, and 67 kd, with minor labeling of 85-kd and 100-kd proteins. In normal cell preparations, only the 37-kd protein was phosphorylated. The 37-kd, 50-kd, and 67-kd proteins have been characterized only with respect to phosphoamino acid content: each contains predominantly phosphotyrosine. These phosphorylations are thought to not be $pp60^{src}$ mediated, since they were unaffected by preincubation of vesicles with TBR serum, which bound $pp60^{src}$ and would have inhibited phosphorylation of $pp60^{src}$ substrates. Thus, these proteins are unlikely to be $pp60^{src}$ substrates.

GARBER et al. (1983a) have described a tyrosyl kinase activity attributable to $pp60^{src}$ that is associated with membrane vesicles prepared from RSV-infected CEF cells. This activity phosphorylates $pp60^{src}$ and several other proteins whose molecular weights are 68 kd, 90 kd, and 130 kd by analysis on SDS-PAGE. The membrane vesicles are prepared by a procedure in which the bulk of $pp60^{src}$ remains membrane bound and the tyrosyl protein kinase activity assayed in vitro is predominantly membrane associated. The 60-kd phosphoprotein product of the in vitro reaction was shown to be $pp60^{src}$ on the basis of the following observations: (a) it was immunoprecipitated by TBR serum; (b) it contained phosphotyrosine on the carboxy-terminal 26-kd V8 protease peptide, and (c) its phosphorylation was dependent upon the temperature of growth of the cells in preparations from *ts*NY68-infected cells and upon salt concentration in preparations from rASV 1702-infected cells, in which $pp60^{src}$ membrane association in cell homogenates is salt dependent (KRUEGER et al. 1982). The 130-kd, 90-kd, and 68-kd phosphotyrosine-containing proteins have not been characterized extensively, but phosphorylation of these candidate $pp60^{src}$ substrates parallels the reactivation of $pp60^{src}$ kinase activity in *ts*NY68-infected cells shifted from the restrictive to the permissive temperature. The 130-kd protein may be vinculin. The 68-kd protein may be identical to the 68-kd protein phosphorylated in cytoskeleton preparations. Preliminary partial peptide mapping suggests that the 90-kd protein is different from the 90-kd protein complexed to $pp60^{src}$. The functions of these potential substrates are not known.

9 Concluding Remarks

Abundant evidence has accumulated indicating that $pp60^{src}$ is an integral membrane protein with cytoplasmic orientation. It appears to be synthesized on free polysomes in the cytoplasm and is transported to the plasma membrane, probably by the pp50:pp90 complex. It seems to be concentrated in adhesion plaques and areas of cell-to-cell contact which are specialized regions of the plasma membrane. At least two functional domains of $pp60^{src}$ have been defined: an 8-kd amino-terminal region appears to be involved in membrane binding, and a larger region near the carboxy-terminal end

Table 8. Viral *onc* genes and properties of their protein products

Gene	Virus	Probable animal of origin	Protein product(s)	*gag* Fusion	Tyrosyl protein kinase	Post-translational modification			Subcellular localization		
						PO_4	CHO	lipid	Integral protein in plasma membrane	Peripheral membrane protein or membrane interaction	Other
src	RSV	Chicken	p60	–	+	+	–	+	+		Cytoskeleton association of p60 association as determined by detergent fractionation
	B77	Chicken	p60	–	+	+	–	+	+		
	rASV	Chicken, Quail	p60	–	+	+	–	+	+		
	rASV	Chicken	p62.5, p56	–	+	+	–	–	–	+	May be largely cytosolic
fps	FSV	Chicken	p130, p140	+	+	+	–		–	+ (p130)	May be largely cytosolic
	PRCII	Chicken	p105, p110	+	+	+	–				Cytoskeleton association of p110 as determined by detergent fractionation
	UR1	Chicken	p150	+	+	+	–				
	16L	Chicken	p142	+	+	+	–				
yes	Y73	Chicken	p90	+	+	+					
	ESV	Chicken	p80	+	+	+					
ros	UR2	Chicken	p68	+	+	+					
myb	AMV	Chicken									
	E26	Chicken	p130	+							
erbB	AEV	Chicken	p68	–	–	+	+		(+)		
myc	MC29	Chicken	p110	+	–	+			–	–	Nucleus DNA
	MH2	Chicken	p100	+	–				–	–	Cytosol RNA
	CMII	Chicken	p90	+	–				–	–	Nucleus DNA
	OK10	Chicken	p200	+							
rel	AEV	Turkey									

mos	Mo-MSV	Mouse	p37	–	–	+			–	–	Soluble cytosol
	Mo-MSV	Mouse	p85	+	–	+					
	GZ-MSV	Mouse	p37	–	–	+					
ras	Ki-MSV	Rat	p21	–	–	+		+	(+)		
	Ha-MSV	Rat	p21	–	–	+		+	(+)		
	Ra-Rasv	Rat	p29	–	–	+					
bas	Balb-msv	Mouse	p21								
abl	Ab-muLV	Mouse	p120	+	+	+	–	+	(+)		
fes	ST-FeSV	Cat	p85	+	+	+	+		(+)		
	GA-FeSV	Cat	p110	+	+	+	+		(+)		
fms	MS-FeSV	Cat	p170	+	+	±			–	–	Co-localizes with intermediate filaments
sis	SiSV	Monkey	p28	–	–	–					
fos	FBJ-MSV	Mouse	p55	–	–	+					

Note. (+), probable or likely integral membrane protein based on known properties, though all criteria listed in Table 4 for biochemical definition of membrane integration have not been met contains the kinase activity. Mutations which affect the structure of the membrane binding domain lead to decreased oncogenic potential of the altered pp60^{src}s. This suggests that both structural domains of pp60^{src} may have important functional roles in producing oncogenic transformation of cells. Both the localization of pp60^{src} in the plasma membrane and the identification of numerous putative substrates of pp60^{src} as plasma membrane proteins suggest that pp60^{src} is likely to cause cell transformation by altering important growth regulatory systems which reside in the plasma membrane. Future work is likely to identify other structural regions of pp60^{src} which are important in its function and will hopefully clarify the extent to which structural features of the amino terminus allow pp60^{src} to interact with the membrane and with specific membrane targets.

Table 8 summarizes available data on the localization of the protein products of numerous viral *onc* gene products. Several other *onc* products also appear to be integral membrane proteins. All of those proteins, with the exception of p21^{ras}, have been identified as tyrosyl protein kinases. The interaction with membranes involves fatty acid acylation in the case of the products of the *src*, *ras*, and *abl* genes. It remains to be determined if the membrane association of other *onc* proteins requires fatty acid acylation.

Definition of the subcellular localization of specific retroviral transforming proteins provides a starting point for investigation of growth regulatory functions residing in specific subcellular locations. Since different *onc* proteins interact with many different subcellular sites, it is likely that cellular transformation may be accomplished by affecting different regulatory or growth control pathways which exist in cells. Future work must define whether *onc* proteins which are located in the same subcellular site act by affecting a common target.

10 Appendix: General Methods Used for Subcellular Localization of Proteins

10.1 Cellular Fractionation

Cellular fractionation is one of the most powerful techniques for determining the subcellular localization of a protein. This method entails fractionation of cellular homogenates by differential centrifugation and further separation of intracellular organelles and membranes by equilibrium centrifugation in continuous or discontinuous sucrose gradients (isopyknic separation). An excellent introduction to the history and methods of cellular fractionation of tissues is provided by DE DUVE (1971), and a detailed description of theory and techniques of cellular fractionation has been written by BEAUFAY and AMAR-COSTESEC (1976). Cellular fractionation techniques which have been developed specifically for organelles or membranes derived from tissues will not be discussed in detail in this review. While many different isolation procedures have been developed for organelles or membranes derived from different types of tissue culture cells (GLICK 1976; WARREN and GLICK 1971; SMUCKLER et al. 1976; FLEISCHER and PACKER 1974; NEVILLE 1975), most are preparative procedures and are not suitable for analytical cell fractionation to determine the primary subcellular location of a given protein or antigen. The outline and discussion of cell fractionation techniques which follows concentrates principally on methods which have been used to localize virus-encoded proteins in tissue culture cells.

10.1.1 Subcellular Component Separation by Differential Centrifugation and Isopyknic Centrifugation

The cellular fractionation method developed by HAY (1974) for analysis of cellular membranes in normal and influenza virus-infected chick embryo fibroblasts has been applied to the study of the localization of pp60^{src} (GOLDBERG et al. 1980; KRUEGER et al. 1980a, 1980b; COURTNEIDGE et al. 1980; KRZYZEK et al. 1980; GARBER et al. 1982) and pp60^{src} substrates (COURTNEIDGE et al. 1983; AMINI and KAJI 1983; GREENBERG and EDELMAN 1983b). Because of the similarity of this method to fractionation methods used by others to determine the intracellular location of *onc* gene-encoded proteins (FELDMAN et al. 1983), our discussion of cell fractionation technology will consider chiefly this procedure.

10.1.1.1 Differential Centrifugation

Homogeneous suspensions of tissue culture cells are swollen in hypotonic media, thereby increasing their fragility, and then are disrupted by shear force in a Dounce homogenizer. The degree of cell breakage is monitored by phase contrast microscopy and is optimized to disrupt as many cells as possible while retaining nuclei intact. Optimal lysis cannot usually be obtained by homogenization in Dounce or Potter-Elvehjem tissue grinders when cells are suspended in isotonic media. Cellular disruption is easily achieved by sonication in isotonic media. However, nuclei are totally fragmented and many small nonsedimentable membrane vesicles are created. Thus cellular disruption by sonication is not widely applicable to cell fractionation by differential centrifugation.

Figure 14 outlines the cellular fractionation procedure described by HAY (1974) and includes some minor modifications described by KRUEGER et al. (1980a). Cells grown in monolayer culture are washed with phosphate-buffered saline, harvested by scraping, and resuspended in a hypotonic TRIS buffer. The swollen cells are disrupted by Dounce homogenization and are separated into four fractions by differential centrifugation:

1. A nuclear pellet fraction (N) containing nuclei, all unbroken cells, and a variable amount of contaminating plasma membranes, endoplasmic re-

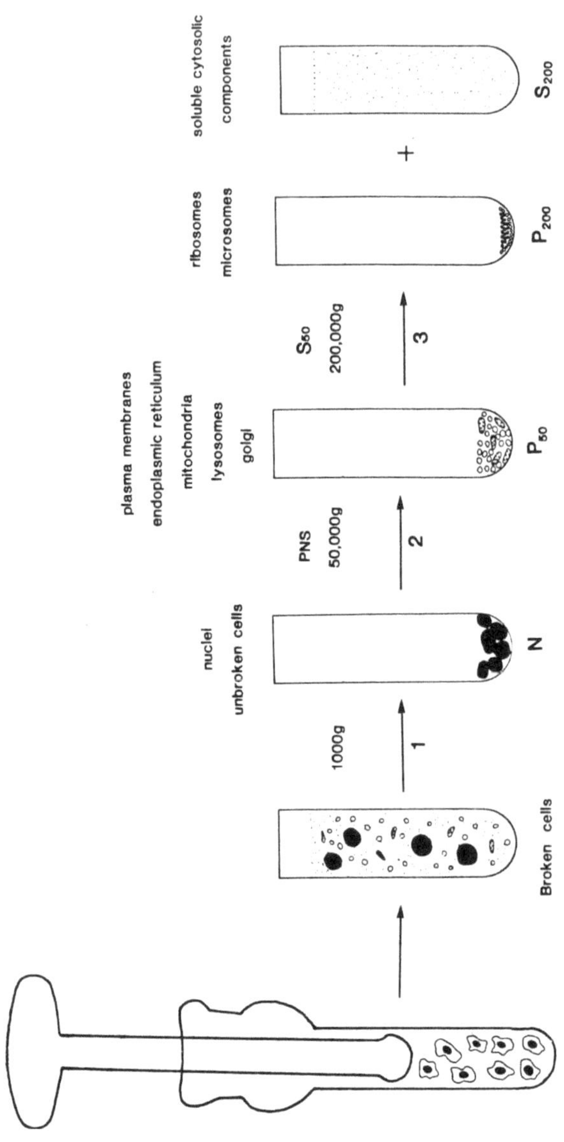

Fig. 14. Scheme of cellular fractionation by differential centrifugation. *1*, *2*, and *3* refer to steps listed in the text. Only the principal subcellular organelles or components in each fraction are listed in the figure

ticulum, and other cellular membranes and organelles is obtained by centrifugation of the cell homogenate for 5 min at 1000 *g*. This fraction contains 10%–15% of the total protein in homogenates from normal or transformed CEFs (HAY 1974; KRUEGER et al. 1980a, 1980b).

2. Further centrifugation of the postnuclear supernatant (PNS) for 20 min at 50000 *g* yields a pellet (designated P_{50}) containing most cellular organelles and membranes and a supernatant fraction. S_{50}, containing small microsomal membrane vesicles, free ribosomes, and soluble cytoplasmic

constituents. The principal subcellular components found in the P_{50} fraction are listed in Fig. 14.

3. Centrifugation of the S_{50} fraction for 60 min at 200000 *g* produces another pellet fraction (P_{200}) containing ribosomes and microsomes and another supernatant fraction (S_{200}) containing soluble cytoplasmic components. It should be noted that the conventional designation of a soluble cytoplasmic fraction is that supernatant resulting from the centrifugation of a cellular homogenate for 60 min at a minimum of 100000 *g*. A soluble cytoplasmic fraction can be prepared directly from a cellular homogenate without intervening slower centrifugation steps or can include several intermediate centrifugation steps.

The Hay cell fractionation procedure was developed and characterized for CEFs and is not directly applicable to the fractionation of mammalian cells. For example, an N fraction obtained from SR-NRK cells contains 32% of cell protein (vs 13% in CEF N fractions) and is more highly contaminated with endoplasmic reticulum and plasma membranes (Courtneidge et al. 1980). Similarly obtained fractions from different cell types may contain quantitatively, if not qualitatively, different intracellular organelles or membranes. Therefore subcellular fractions derived from different cell types should be characterized carefully (see Sect. 10.2) before results are interpreted or comparisons made.

10.1.1.2 Isopyknic Centrifugation

Definitive localization of a protein by differential centrifugation is feasible if that protein is exclusively associated with the nucleus or if it is wholly soluble in the cytoplasm. However, if a protein is associated with cytoplasmic organelles or membranes, additional fractionation steps are necessary for specific localization. These might include further separation of P_{50} (membrane and organelle) fractions on discontinuous sucrose gradients by isopyknic centrifugation (centrifugation of organelles to equilibrium density in sucrose columns). Figure 15 shows the procedure modified from Hay (1974) for isopyknic separation of subcellular membranes and organelles on discontinuous sucrose gradients. A P_{50} pellet of membranes and organelles is resuspended in 45% sucrose and layered over a 60% sucrose cushion. The gradient is then successively overlayed with 35%, 30%, 25%, 20%, and 0% sucrose solutions in TRIS-containing buffers. The discontinuous sucrose gradient is centrifuged for 16 h at 80000 *g*. During this centrifugation, membranes and organelles redistribute in the gradient until they achieve their isopyknic density. Material which bands at each of the sucrose interfaces is collected with a pipette or syringe, diluted with buffer solution, and recentrifuged at 150000 *g* to concentrate the material in a pellet which can be resuspended in a small volume for subsequent analysis. The material at each of the five interfaces (listed in Fig. 15) has been characterized extensively by biochemical or electron microscopic techniques (Hay 1974; Krueger et al. 1980a). The upper two fractions were found to contain

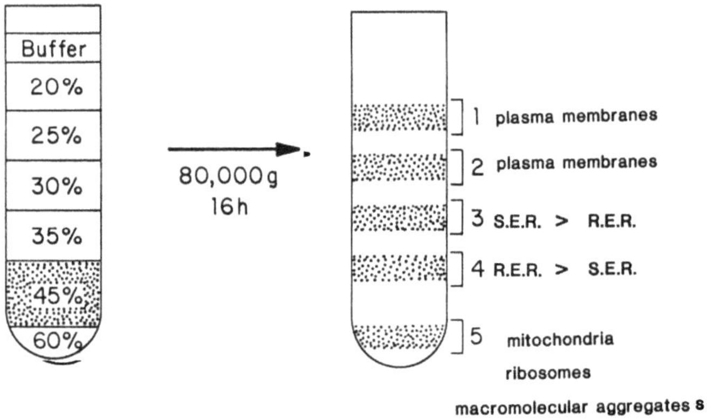

Fig. 15. Fractionation of subcellular membranes and organelles by isopyknic centrifugation. A P_{50} pellet from differential centrifugation is resuspended in a pH 7.0 TRIS buffer containing 45% w/w sucrose and layered over a 60% sucrose cushion. The 45% sucrose solution containing the subcellular components is then successively overlayed with pH 7.0 TRIS buffer containing decreasing percentages of sucrose and finally with a buffer solution. The resulting discontinuous sucrose gradient (*left*) is centrifuged for 16 h at 80000 g. Membrane vesicles and subcellular organelles redistribute to interface regions between the different density sucrose layers according to their isopyknic density. The interface regions (labeled *1–5*, *right*) are enriched in specific membranes and organelles as indicated above. *S.E.R.*, smooth endoplasmic reticulum; *R.E.R.*, rough endoplasmic reticulum

smooth membranous vesicles enriched for 5′-nucleotidase, an enzyme found exclusively in association with plasma membranes. These fractions contained small amounts of activity of NADH diaphorase, an enzyme found in association with endoplasmic reticulum. Since fractions 3 and 4 contained the bulk of NADH diaphorase activity and morphologically verifiable endoplasmic reticulum, it is assumed that fractions 1 and 2 are specifically enriched in cellular plasma membranes. Mitochondria and free ribosomes were found chiefly in fraction 5, which also contained aggregated macromolecular material. These fractions, however, were not evaluated for the presence of enzymes specific for lysosomes or Golgi vesicles. It can be assumed that these membrane vesicles, which are minor intracellular components in cultured fibroblasts, would be quantitatively minor contaminants in fractions 1 and 2 since their density is similar to plasma membranes (Beaufay and Amar-Costesec 1976). Thus separation of P_{50} membrane-organelle fractions in discontinuous sucrose gradients allows recovery and analysis of fractions enriched in specific cellular components. A similar fractionation procedure can be applied to cells derived from other species, but the precise membrane-organelle content of each fraction should be carefully defined by biochemical and microscopic procedures.

10.1.2 Caveats Concerning Cell Homogenization

In order to fairly compare subcellular distribution of a protein in different types of tissue culture cells, it is essential that a similar percentage of cells be broken. Efficient disruption of a population of cells by Dounce homogenization depends on the fragility of the cell (a function of both cell type and buffer composition) and the shear force generated by the individual Dounce homogenizer being used. Our experience indicates that Dounce homogenizers with tight-fitting pestles supplied by different manufacturers and even different-sized homogenizers supplied by the same manufacturer differ in their ability to disrupt cells. For example, the transformed rat cell RR1022 is efficiently disrupted only by an extremely tight-fitting Dounce homogenizer, while normal and transformed CEFs suspended in exactly the same hypotonic buffer are much more easily disrupted.

It should be emphasized that efficient cell disruption by shear forces in a homogenizer depends largely on the specific configuration of the type of homogenizer. A Dounce homogenizer contains a smooth glass barrel with a smooth, tight-fitting glass plunger. Other types of homogenizers may resemble the Dounce homogenizer, but may produce substantially different homogenization results. These include homogenizers with Teflon plungers, with abrasive glass surfaces on either the barrel or plunger, or of a different configuration than that shown in Fig. 14. Sometimes it is desirable to break cells in a small volume of buffer solution to facilitate subsequent analysis of fractions. A motor-driven Potter-Elvehjem homogenizer is preferred in such a case, as small-volume Dounce homogenizers with sufficiently tight-fitting pestles are difficult to obtain. This type of homogenization is likely to produce smaller membrane vesicles and therefore necessitate longer centrifugation times or higher *g* forces to sediment all membranes or organelles.

All fractionation procedures subsequent to cell disruption involve rate-dependent, differential sedimentation of subcellular components by centrifugation. Differential centrifugation steps are ineffective unless subcellular organelles remain intact and factors affecting size and shape of membrane sheets and organelles are controlled. Both aspects are affected by the composition of the lysis media when Dounce homogenization is employed to disrupt cells. CEFs may be disrupted in media containing either divalent cation chelators (e.g., EDTA) or Mg^{+2}. Most mammalian cell nuclei are lysed unless Mg^{+2} is added to the homogenization media. Disruption of nuclei may lead either to contamination of other fractions with nuclear debris, or to increased viscosity of the lysate, thereby preventing subsequent purification of other subcellular components. Such damage also may result in artifactual association of proteins with DNA. Inclusion of EDTA or other divalent cation chelators causes the removal of ribosomes from endoplasmic reticulum and may thus affect the fractionation of both ribosomes and endoplasmic reticulum. The size and configuration of plasma membranes obtained by Dounce homogenization is also affected by homogenization media composition. For example, exposure of undisrupted cells to Zn^{+2}, TRIS base, acid pH (<6.0), or fluorescein mercuric acetate (FMA) increases the stability of plasma membranes and affects the size of membrane sheets and vesicles obtained after Dounce homogenization (Warren et al. 1966). Large sheets of plasma membranes obtained from Dounce-homogenized CEF cells exposed to TRIS buffers cosediment with nuclei in differential centrifugation (Hay 1974).

10.1.3 Analysis of Information Derived from Cellular Fractionation

The precision with which a protein can be localized in a particular subcellular location depends on a number of factors, including the specificity of interaction of a protein with a specific organelle, the method of cell disruption, and the degree to which particular cellular organelles are contaminated with other organelles. Organelle fractionation is always a compromise between high yield and high purity. There are various limitations and restrictions that should be considered in the analysis of cell fractionation data.

Proteins have been shown to redistribute into different compartments during cell homogenization or fractionation. Both the ionic and chemical composition of the homogenization media can affect the fractionation of peripheral membrane proteins so that they distribute either with membranes or soluble cytoplasmic proteins (see SINGER 1974). Similarly, lysis of cells in low osmotic strength buffers may cause soluble cytoplasmic proteins to associate with membranes through ionic bonds. For example, EMMELOT and BOS (1966) showed that triose phosphate dehydrogenase artifactually associated with membranes when rat hepatoma cells were extracted in low-salt buffer. Several reports have documented the extraction of integral membrane proteins into soluble fractions (see SINGER 1974). Specific controls can address these issues (GLICK 1976; SMUCKLER et al. 1976; EMMELOT and BOS 1966).

Rigorously pure fractions of specific subcellular organelles are rarely obtained:

Nuclear Fractions. Nuclear fractions are frequently contaminated with plasma membrane sheets and to a variable degree with endoplasmic reticulum, Golgi, and lysosomal membranes. In order to determine if a particular protein is truly localized in the nucleus or is artifactually associated with membranes or whole cells that are contaminating a nuclear fraction, the following precautionary steps might be carried out: (a) nonionic detergent preparation of nuclei, (b) modified sucrose gradient centrifugation steps (HAY 1974), (c) enzyme marker studies, and (d) light and electron microscopic characterization of the contents of this fraction.

Membranous or Organelle Fractions. Techniques that have been devised for organelle purification from tissues where one starts with large quantities (greater than 1 g) of material, are not appropriately applied to tissue culture systems. The techniques which have been applied to localization of virus-encoded proteins in tissue culture systems have been designed for good separation of organelles and membranes and maximal recovery of starting material. The cell organelle separation procedures described in Sect. 10.1.1 are rapid, produce high yields, and can be used for both analytical and preparative separation of organelles and membranes. Step gradients, however, only give fractions enriched in specific organelles or membranes. Quantitative assessment of association of a particular protein with specific membranes is based on the fractionation pattern relative to marker proteins or enzymes and on the specific enrichment of a protein in a fraction. Thus the type of separation achieved is neither quantitatively nor qualitatively perfect. A major limitation of isopyknic separation techniques is that different subcellular organelles or membranes with similar densities cofractionate. A recurring problem in purification of plasma membranes from cells by isopyknic sucrose gradient centrifugation has been coseparation of plasma membranes, lysosomes, and Golgi membranes. Some difficulty in separation of lysosomal and plasma membranes might be anticipated, since lysosomal membranes are formed in part from pinocytotic or phagocytotic vesicles derived from plasma membranes. The lipid composition of plasma membranes, lysosomal membranes, and Golgi membranes is similar (ROBINSON 1975; QUINN 1976). Plasma membranes, Golgi membranes, and lysosomes derived from liver cells frequently cofractionate in isopyknic separation systems (BEAUFAY and AMAR-COSTESEC 1976; DE DUVE 1971) and cofractionation may be anticipated for these membranes derived from tissue culture. KRZYZEK et al. (1980) fractionated cellular organelles and membranes from in vitro cultured field vole cells and monitored the purification of plasma membranes by ^{125}I-labeled wheat germ agglutinin binding and the purification of lysosomes by measuring acid phosphatase activity. The separation pattern of these two membranous components was similar, if not identical, in isopyknic sucrose gradients. Similar cofractionation of lysosomal, Golgi, and plasma membranes might be expected in other types of tissue culture cells. It should be evident that one cannot use isopyknic centrifugation techniques to distinguish unambiguously between association of a protein with plasma membranes and with Golgi or lysosomal membranes. Reviews by GLICK (1976) and NEVILLE (1975) discuss more extensively the procedural artifacts and the methods used to establish purity of specific subcellular fractions.

10.2 Microscopic Techniques for Subcellular Localization of Proteins

Immunofluorescence and immunoelectron microscopy allow the investigator to study the intracellular location of a protein or antigen in fixed, undisrupted cells. These techniques are useful both as primary methods of localizing an antigen and as secondary procedures for the confirmation of the results of subcellular fractionation. This section will discuss only techniques which have been applied to the localization of pp60^{src} and other virus-encoded transforming proteins. The reader is referred to DE PETRIS (1978) for an extensive description of the theory and techniques of immunofluorescence and immunoelectron microscopy.

10.2.1 Immunofluorescence Microscopy

To localize proteins by indirect immunofluorescence techniques, cells are grown on a glass or plastic surface and are fixed with formaldehyde or acetone in order to maintain the in situ location of cellular components. Fixed cells are made permeable to antibody by subsequent treatment with detergents. Cells are then incubated with a primary antibody directed against the antigen of interest, and after a suitable incubation period the cells are extensively washed in buffered saline to remove nonreacting immunoglobulins. A second antibody, conjugated with a fluorescent dye such as rhodamine or fluorescein, directed against the primary antibody, is then allowed to react with the cells; excess, nonreacting secondary antibody is removed by a series of washes. The pattern of immunofluorescence can then be visualized with a microscope equipped with epifluorescence optics. This technique provides useful information about the subcellular localization of a protein provided several criteria are satisfied: (a) cells must be fixed so that the antigen remains in situ during permeabilization and washing steps, (b) the antigen must not be denatured during fixation or permeabilization beyond the ability of the primary antibody to specifically react with it, (c) the specificity of the primary antibody must be determined, and (d) intracellular structures must be identifiable at the light-microscopic level.

Some difficulties are encountered in using indirect immunofluorescence techniques. Fixation procedures for immunofluorescence do not always preserve the integrity of intracellular organelles to the same extent as fixation procedures used for electron microscopy. Thus, for instance, the final disposition of a soluble, cytoplasmic antigen after fixation cannot be predicted with certainty. Acetone fixation sometimes denatures antigens, thereby preventing recognition by the primary antibody. Alternatively, the antigens may be insufficiently fixed and would therefore be removed during the washing steps. The third criterion of demonstrated antibody specificity is probably the most difficult to satisfy. The most straightforward demonstration of antibody specificity for a specific protein is direct immunoprecipitation followed by identification of the immunoprecipitated protein on SDS-polyacrylamide gels. A further complication of indirect immunofluorescence is that unfractionated primary antisera may contain nonprecipitating antibodies to other proteins or to other nonprotein cellular antigens (e.g., nucleic acids or glycolipid complexes). Even in instances where monospecific antibodies are prepared by affinity chromatography with a pure antigen, numerous cellular proteins may be recognized because of common antigenic sites (NIGG et al. 1982b). Finally, not all subcellular components can be identified by light microscopy. In practice, those antigens associated with nuclei, cytoskeletal elements, and plasma membranes are easiest to visualize by immunofluorescence; those antigens associated with internal organelles, membranes, or the soluble cytosol are at best tenuously identified, since the integrity of these structures is often destroyed by conventional fixation/permeabilization procedures (WILLINGHAM et al. 1978, 1979).

10.2.2 Immunoelectron Microscopy

WILLINGHAM et al. (1979, 1980) have used the EGS (ethyldimethyl aminopropyl carbodiimide/glutaraldehyde/saponin) procedure to localize the

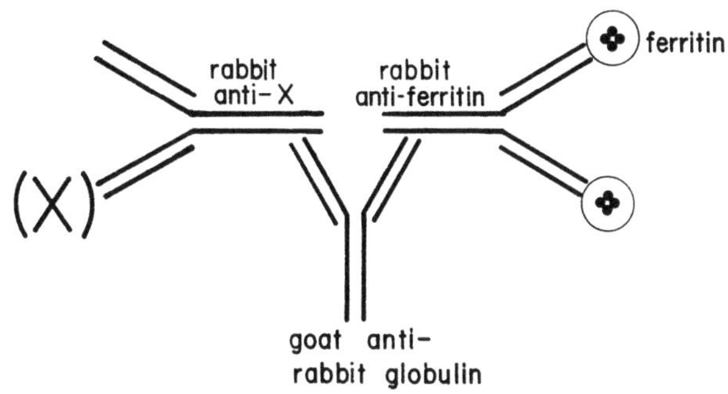

Fig. 16. A four-step antibody "sandwich" which is used to localize protein "X" within cells by immunoelectron microscopy

transforming proteins encoded by RSV and Harvey murine sarcoma virus. This procedure allows better preservation of cellular membranes and membrane-limited organelles than do the fixation and permeabilization procedures used for immunofluorescence microscopy. Cells are fixed with the protein cross-linking reagents ethyldimethyl aminopropyl carbodiimide and glutaraldehyde for a defined period of time and then are made permeable to antibodies by treatment with saponin. A series of antibodies are incubated with the cells to generate an "antibody bridge," which is shown in Fig. 16. The conjugated ferritin cores are detected by electron microscopy, enhancing the resolution in localization of an antigen. The specificity of localization data using this technique is limited by the same criteria as immunofluorescence microscopy, e.g., cell structures must be fixed, antigens must remain in their original position during manipulations, and the antibody specificity must be carefully documented. This technique is widely applicable to subcellular localization of proteins, provided that the primary antibody has demonstrated specificity (WILLINGHAM et al. 1978).

10.3 Cellular Fractionation with Nonionic Detergents

Numerous reports have described the fractionation of cells into "cytoskeletal" and "soluble" compartments following extraction with nonionic detergents (BURR et al. 1980, 1981; CHENG and CHEN 1981a, b; COOPER and HUNTER 1982; SCHAFFHAUSEN et al. 1982; GEIGER 1982). These fractionation procedures were developed to study spatial relationships among the various cytoskeletal components and to probe cytoskeletal elements with antibodies or other reagents (BROWN et al. 1976; TROTTER et al. 1978; HENDERSON and WEBER 1979; OSBORN and WEBER 1977; WEBSTER et al. 1978). After extraction of monolayer cultures with nonionic detergents, a "cytoskeletal" residue, enriched in actin, vimentin, and a 230-kd protein, remains on the dish. Nuclei are also retained in the "cytoskeletal framework" residue. Characterization of this cytoskeletal compartment by indirect immunofluorescence microscopy with anti-actin antibodies (OSBORN and WEBER 1977;

Henderson and Weber 1979) has established that actin-containing fibers are present in the cytoskeletal residue. Microtubules are not usually retained in such procedures (Henderson and Weber 1979). Cytoskeleton residues prepared by different investigators appear to be reasonably similar when visualized by electron microscopy (Brown et al. 1976; Henderson and Weber 1979; Trotter et al. 1978; Webster et al. 1978; Schliwa et al. 1981). However, characterization of these cytoskeletal fractions by one- and two-dimensional gel electrophoresis has revealed wide differences in the range of proteins contained in these preparations. Cytoskeletons produced by Triton X-100 extraction of CEFs showed only three major residual proteins (Brown et al. 1976), while cytoskeletons prepared from BSC-1 cells using the same Triton concentration showed more than 200 proteins (Schliwa et al. 1981). It is likely that specific differences in ionic and pH composition of the extraction buffer, as well as in the type of cell used, may account in large part for these differences. However, even in the best-characterized extraction procedures, little attention has been paid to the final disposition of verified membrane proteins and of soluble cytosolic proteins. Schliwa et al. (1981) showed that only 14 plasma membrane proteins were extracted from BSC-1 cells by Brij-58, while the remaining plasma membrane proteins were found to be attached to residual cytoskeleton. It should be pointed out that these cytoskeleton fractionation procedures cannot distinguish between a location of a protein in cellular membranes or in the soluble phase of the cytosol, since some integral membrane proteins are extracted by low concentrations of nonionic detergents, while others are retained (Yu et al. 1973; Ben-Ze'ev et al. 1979). Unless experimental data verify the disposition of soluble cytoplasmic proteins and known membrane proteins, cytoskeletons prepared by nonionic detergent extraction of cells should not be used to localize proteins to specific subcellular compartments. However, once the intracellular location of a protein is established by other means, such as cell fractionation important biochemical and structural information can be attained by treatment of cells or subcellular fractions with nonionic detergents (Yu et al. 1973).

Acknowledgements. We thank Dr. Hidesaburo Hanafusa and Dr. Igor Tamm for critical reading of the manuscript. Work in the laboratory of A.R.G. has been supported by grants from the National Institutes of Health.

References

Abercrombie M, Heaysman JEM, Pegrum SM (1971) The locomotion of fibroblasts in culture. IV. Electron microscopy of the leading lamella. Exp Cell Res 67:359–367

Amini S, Kaji A (1983) Association of pp36, a phosphorylated form of the presumed target protein for the *src* protein of Rous sarcoma virus, with the membrane of chicken cells transformed by Rous sarcoma virus. Proc Natl Acad Sci USA 80:960–964

Anderson DD, Beckmann RP, Harms EH, Nakamura K, Weber MJ (1981) Biological properties of "partial" transformation mutants of Rous sarcoma virus and characterization of their pp60^{src} kinase. J Virol 37:445–458

Antler A, Greenberg ME, Goldberg AR, Edelman GM, Hanafusa H (to be published) Transformation by avian sarcoma viruses and the distribution of vinculin and actin in transformed cells: the role of vinculin phosphorylation

Barnekow A, Bauer H, Boschek CB, Friis RR, Ziemiecki A (1981) Rous sarcoma virus transformation: action of the *src* gene product. In: Schweiger H (ed) International cell biology 1980–1981. Springer, Berlin, pp 457–466

Barnekow A, Boschek CB, Ziemiecki A, Friis RR, Bauer H (1982) Demonstration of the Rous sarcoma virus pp60^{src} and its associated protein kinase on the surface of intact cells. In: Winnacker E, Schoene H-H (eds) Genes and tumor genes. Raven, New York, pp 65–73

Bauer H, Barnekow A, Rose G (1982) The transforming protein of Rous sarcoma virus pp60^{src}: growth and cell proliferation inducing properties. In: Dumont JE, Nunez J, Schultz G (eds) Hormones and cell regulation, vol 6. Elsevier, New York, pp 187–205

Beaufay H, Amar-Costesec A (1976) Cell fractionation techniques. In: Korn ED (ed) Methods in membrane biology, vol 6. Plenum, New York, pp 1–100

Beemon K, Hunter T (1978) Characterization of Rous sarcoma virus *src* gene products synthesized in vitro. J Virol 28:551–566

Beemon K, Ryden T, McNelly EA (1982) Transformation by avian sarcoma viruses leads to phosphorylation of multiple cellular proteins on tyrosine residues. J Virol 42:742–747

Ben-Ze'ev A, Duerr A, Solomon F, Penman S (1979) The outer boundary of the cytoskeleton: a lamina derived from plasma membrane proteins. Cell 17:859–865

Birchmeier W, Liberman TA, Imhof BA, Kreis TE (1982) Intracellular and extracellular components involved in the formation of ventral cell surfaces of fibroblasts. Cold Spring Harbor Symp Quant Biol 46:755–768

Bishop JM, Courtneidge SA, Levinson AD, Oppermann H, Quintrell N, Sheiness DK, Weiss SR, Varmus HE (1980) Origin and function of avian retrovirus transforming genes. Cold Spring Harbor Symp Quant Biol 44:919–930

Blithe DB, Richert ND, Pastan IH (1982) Purification of a tyrosine-specific protein kinase from Rous sarcoma virus-induced rat tumor. J Biol Chem 257:7135–7142

Blobel G (1980) Intracellular protein topogenesis. Proc Natl Acad Sci USA 77:1496–1500

Brown S, Levinson W, Spudich JA (1976) Cytoskeletal elements of chick embryo fibroblasts revealed by detergent extraction. J Supramol Struct Cell Biochem 5:119–130

Brugge JS, Erikson RL (1977) Identification of a transformation-specific antigen induced by an avian sarcoma virus. Nature 269:346–348

Brugge JS, Darrow D (1982) Rous sarcoma virus-induced phosphorylation of a 50,000-molecular weight cellular protein. Nature 295:250–253

Brugge JS, Steinbaugh PJ, Erikson RL (1978) Characterization of the avian sarcoma virus protein p60^{src}. Virology 91:130–140

Brugge JS, Erikson E, Erikson RL (1981) The specific interaction of the Rous sarcoma virus transforming protein, pp60^{src}, with two cellular proteins. Cell 25:363–372

Brugge JS, Yonemoto W, Darrow D (1983) Interaction between the Rous sarcoma virus transforming protein and two cellular phosphoproteins: analysis of the turnover and distribution of this complex. Mol Cell Biol 3:9–19

Bryant D, Parsons JT (1982) Site-directed mutagenesis of the *src* gene of Rous sarcoma virus: construction and characterization of a deletion mutant temperature sensitive for transformation. J Virol 44:683–691

Bunte T, Owada MK, Donner P, Boschek CB, Moelling K (1981) Association of the transformation-specific protein pp60^{src} with the membrane of an avian sarcoma virus. J Virol 38:1034–1047

Burr JG, Dreyfuss G, Penman S, Buchanan JM (1980) Association of the *src* gene product of Rous sarcoma virus with cytoskeletal structures of chicken embryo fibroblasts. Proc Natl Acad Sci USA 77:3484–3488

Burr JG, Lee SR, Buchanan JM (1981) In situ phosphorylation of proteins associated with the cytoskeleton of chick embryo fibroblasts. Cold Spring Harbor Conf Cell Prolif 8:1217–1232

Carr SA, Biemann K, Shoji S, Parmalee DC, Titani K (1982) *n*-Tetradecanoyl is the NH_2-terminal blocking group of the catalytic subunit of cyclic AMP-dependent protein kinase from bovine cardiac muscle. Proc Natl Acad Sci USA 79:6128–6131

Cheng YSE, Chen LB (1981a) Detection of phosphotyrosine-containing 34,000-dalton protein in the framework of cells transformed with Rous sarcoma virus. Proc Natl Acad Sci USA 78:2388–2392

Cheng YSE, Chen LB (1981b) Alterations in protein phosphorylation in cells transformed by Rous sarcoma virus. Cold Spring Harbor Conf Cell Prolif 8:1233–1246

Collett MS, Erikson RL (1978) Protein kinase activity associated with the avian sarcoma virus *src* gene product. Proc Natl Acad Sci USA 75:2021–2024

Collett MS, Purchio AF, Erikson RL (1980) Avian sarcoma virus transforming protein, pp60^{src}, shows protein kinase activity specific for tyrosine. Nature 285:167–169

Cooper JA, Hunter T (1981a) Changes in protein phosphorylation in Rous sarcoma virus-transformed chicken embryo cells. Mol Cell Biol 1:165–178

Cooper JA, Hunter T (1981b) Four different classes of retroviruses induce phosphorylation of tyrosines present in similar cellular proteins. Mol Cell Biol 1:394–407

Cooper JA, Hunter T (1981c) Similarities and differences between the effects of epidermal growth factor and Rous sarcoma virus. J Cell Biol 91:878–883

Cooper JA, Hunter T (1982) Discrete primary locations of a tyrosine protein kinase and of three proteins that contain phosphotyrosine in virally transformed chick fibroblasts. J Cell Biol 94:287–296

Cooper JA, Hunter T (1983) Identification and characterization of cellular targets for tyrosine protein kinases. J Biol Chem 258:1108–1115

Cooper JA, Hunter T (to be published) Analysis of substrates for tyrosine protein kinases. In: Squaw Valley Symposium: tumor viruses and differentiation

Cooper JA, Nakamura KD, Hunter T, Weber MJ (1983a) Phosphotyrosine-containing proteins and expression of transformation parameters in cells infected with partial transformation mutants of Rous sarcoma virus. J Virol 46:15–28

Cooper JA, Reiss NA, Schwartz RJ, Hunter T (1983b) Three glycolytic enzymes are phosphorylated at tyrosine in cells transformed by Rous sarcoma virus. Nature 302:218–223

Courtneidge SA, Bishop JM (1982) The transit of pp60$^{v\text{-}src}$ to the plasma membrane. Proc Natl Acad Sci USA 79:7117–7121

Courtneidge SA, Levinson AD, Bishop JM (1980) The protein encoded by the transforming gene of avian sarcoma virus (pp60^{src}) and a homologous protein in normal cells (pp60$^{proto\text{-}src}$) are associated with the plasma membrane. Proc Natl Acad Sci USA 77:3783–3787

Courtneidge SA, Ralston R, Alitalo K, Bishop JM (1983) Subcellular location of an abundant substrate (p36) for tyrosine-specific protein kinases. Mol Cell Biol 3:340–350

Curtis ASG (1964) The mechanism of adhesion of cells to glass. A study by interference reflection microscopy. J Cell Biol 20:199–215

David-Pfeuty T, Singer SJ (1980) Altered distributions of cytoskeletal proteins vinculin and alpha-actinin in cultured fibroblasts transformed by Rous sarcoma virus. Proc Natl Acad Sci USA 77:6687–6691

de Duve C (1971) Tissue fractionation. J Cell Biol 50:20D–55D

de Petris S (1978) Immunoelectron microscopy and immunofluorescence in membrane biology. In: Korn ED (ed) Methods in membrane biology, vol 9. Plenum, New York, pp 1–201

Dunn GA, Heath JP (1976) A new hypothesis of contact guidance in tissue cells. Exp Cell Res 101:1–14

Emmelot P, Bos JC (1966) Differences in the association of two glycolytic enzymes with plasma membranes isolated from rat liver and hepatoma. Biochim Biophys Acta 121:434–436

Erikson E, Erikson RL (1980) Identification of a cellular protein substrate phosphorylated by the avian sarcoma virus-transforming gene product. Cell 21:829–836

Erikson E, Collett MS, Erikson RL (1978) In vitro synthesis of a functional avian sarcoma virus transforming gene product. Nature 274:919–921

Erikson E, Cook R, Miller GJ, Erikson RL (1981a) The same normal cell protein is phosphorylated after transformation by avian sarcoma viruses with unrelated transforming genes. Mol Cell Biol 1:43–50

Erikson E, Shealy DJ, Erikson RL (1981b) Evidence that viral transforming gene products and epidermal growth factor stimulate phosphorylation of the same cellular protein with similar specificity. J Biol Chem 256:11381–11384

Erikson RL (1981) The transforming protein of avian sarcoma viruses and its homologue in normal cells. Curr Top Microbiol Immunol 91:25–40

Erikson RL, Collett MS, Erikson E, Purchio AF (1979) Evidence that the avian sarcoma virus transforming gene product is a cyclic AMP-independent protein kinase. Proc Natl Acad Sci USA 76:6260–6264

Erikson RL, Collett MS, Erikson E, Purchio AF, Brugge JS (1980) Protein phosphorylation mediated by partially purified avian sarcoma virus transforming-gene product. Cold Spring Harbor Symp Quant Biol 44:907–917

Feldman RA, Wang E, Hanafusa H (1983) Cytoplasmic localization of the transforming protein of Fujinami sarcoma virus: salt-sensitive association with subcellular components. J Virol 45:782–791

Fleischer S, Packer L (1974) Biomembranes (part A). Methods Enzymology, vol 31. Academic, New York

Gallis B, Bornstein P, Brautigan DL (1981) Tyrosylprotein kinase and phosphatase activities in membrane vesicles from normal and Rous sarcoma virus-transformed rat cells. Proc Natl Acad Sci USA 78:6689–6693

Garber EA, Krueger JG, Goldberg AR (1982) Novel localization of pp60^{src} in Rous sarcoma virus-transformed rat and goat cells and in chicken cells transformed by viruses rescued from these mammalian cells. Virology 118:419–429

Garber EA, Krueger JG, Hanafusa H, Goldberg AR (1983a) Temperature-sensitive membrane association of pp60^{src} in tsNY68-infected cells correlates with increased tyrosine phosphorylation of membrane-associated proteins. Virology 126:73–86

Garber EA, Krueger JG, Hanafusa H, Goldberg AR (1983b) Only membrane-associated RSV *src* proteins have amino-terminally bound lipid. Nature 302:161–163

Geiger B (1979) A 130K protein from chicken gizzard: its localization at the termini of microfilament bundles in cultured cells. Cell 18:193–205

Geiger B (1982) Microheterogeneity of avian and mammalian vinculin: distinctive subcellular distribution of different isovinculins. J Mol Biol 159:685–701

Geiger B, Tokuyasu KT, Dutton AH, Singer SJ (1980) Vinculin, an intracellular protein localized at specialized sites where actin microfilament bundles terminate at cell membranes. Proc Natl Acad Sci USA 77:4127–4131

Gilmer TM, Erikson RL (1981) Rous sarcoma virus transforming protein, p60^{src}, expressed in *E. coli*, functions as a protein kinase. Nature 294:771–773

Gilmore T, Radke K, Martin GS (1982) Tyrosine phosphorylation of a 50k cellular polypeptide associated with the Rous sarcoma virus-transforming protein, pp60^{src}. Mol Cell Biol 2:199–206

Glick MC (1975) Isolation of surface membranes from mammalian cells. In: Jamieson GA, Robinson DM (eds) Mammalian cell membranes, vol 1. Butterworth, Boston, pp 47–71

Goldberg AR, Krueger JG, Wang E (1980) Localization and characterization of the *src*-gene product of Rous sarcoma virus. Cold Spring Harbor Symp Quant Biol 44:991–1005

Graziani Y, Erikson E, Erikson RL (1983) Evidence that the Rous sarcoma virus transforming gene product is associated with glycerol kinase activity. J Biol Chem 258:2126–2129

Greenberg ME, Edelman GM (1983a) Comparison of the 34,000 dalton pp60^{src} substrate and a 38,000 dalton phosphoprotein identified by monoclonal antibodies. J Biol Chem 258:8497–8502

Greenberg ME, Edelman GM (1983b) The 34kD-pp60^{src} substrate is located at the inner face of the plasma membrane. Cell 33:767–779

Hanafusa H (1977) Cell transformation by RNA tumor viruses. In: Fraenkel-Conrat H, Wagner RR (eds) Comprehensive virology, vol 10. Plenum, New York, pp 401–483

Hanafusa H, Halpern CC, Buchhagen DL, Kawai S (1977) Recovery of avian sarcoma virus from tumors induced by transformation-defective mutants. J Exp Med 146:1735–1747

Hay AJ (1974) Studies on the formation of the influenza virus envelope. Virology 60:398–418

Heath JP, Dunn GA (1978) Cell to substratum contacts of chick fibroblasts and their relation to the microfilament system. A correlated interference-reflection and high-voltage electron-microscope study. J Cell Sci 29:197–212

Heaysman JEM, Pegrum SM (1973) Early contacts between fibroblasts. Exp Cell Res 78:71–78

Henderson D, Weber K (1979) Three-dimensional organization of microfilaments and microtubules in the cytoskeleton. Exp Cell Res 124:301–316

Henderson LE, Krutzsch HC, Oroszlan S (1983) Myristyl amino-terminal acylation of murine retrovirus proteins: an unusual post-translational protein modification. Proc Natl Acad Sci USA 80:339–343
Hunter T (1980) Proteins phosphorylated by the RSV transforming function. Cell 22:647–648
Hunter T, Sefton BM (1980) The transforming gene product of Rous sarcoma virus phosphorylates tyrosine. Proc Natl Acad Sci USA 77:1311–1315
Hunter T, Cooper JA (1981) Epidermal growth factor induces rapid tyrosine phosphorylation of proteins in A431 human tumor cells. Cell 24:741–752
Hunter T, Sefton BM, Cooper JA (1981) Phosphorylation of tyrosine: its importance in viral transformation and normal cell metabolism. Cold Spring Harbor Conf Cell Prolif 8:1189–1202
Hynes R (1982) Phosphorylation of vinculin by pp60^{src}: what might it mean? Cell 28:437–438
Hynes RO, Destree AT (1978) Relationships between fibronectin (LETS) protein and actin. Cell 15:875–886
Ito S, Richert N, Pastan I (1982) Phospholipids stimulate phosphorylation of vinculin by the tyrosine-specific protein kinase of Rous sarcoma virus. Proc Natl Acad Sci USA 79:4628–4631
Izzard CS, Lochner LR (1976) Cell-to-substrate contacts in living fibroblasts: an interference reflexion study with an evaluation of the technique. J Cell Sci 21:129–159
Kahn P, Nakamura K, Shin S, Smith RE, Weber MJ (1982) Tumorigenicity of partial transformation mutants of Rous sarcoma virus. J Virol 42:602–611
Kamine J, Buchanan JM (1978) Processing of 60,000 dalton *sarc* gene protein synthesized by cell-free translation. Proc Natl Acad Sci USA 75:4399–4403
Kamine J, Burr JG, Buchanan JM (1980) In vitro synthesis and processing of Rous sarcoma virus *src*-gene products. Cold Spring Harbor Symp Quant Biol 44:943–948
Karess RE, Hanafusa H (1981) Viral and cellular *src* genes contribute to the structure of recovered avian sarcoma virus transforming protein. Cell 24:155–164
Kawai S, Hanafusa H (1971) The effect of temperature on the transformed state of cells infected with a Rous sarcoma virus mutant. Virology 46:470–479
Kelley PM, Schlesinger MJ (1982) Antibodies to two major chicken heat shock proteins cross-react with similar proteins in widely divergent species. Mol Cell Biol 2:267–274
Kobayashi N, Kaji AJ (1980) Phosphoprotein associated with activation of the *src* gene product in myogenic cells. Biochem Biophys Res Comm 93:278–284
Kobayashi N, Tanaka A, Kaji A (1981) In vitro phosphorylation of the 36 k protein in extracts from Rous sarcoma virus-transformed chicken fibroblasts. J Biol Chem 256:3053–3058
Krebs EG, Beavo JA (1979) Phosphorylation-dephosphorylation of enzymes. Ann Rev Biochem 48:923–959
Krueger JG, Wang E, Goldberg AR (1980a) Evidence that the *src* gene product of Rous sarcoma virus is membrane associated. Virology 101:25–40
Krueger JG, Wang E, Garber EA, Goldberg AR (1980b) Differences in intracellular location of pp60^{src} in rat and chicken cells transformed by Rous sarcoma virus. Proc Natl Acad Sci USA 77:4142–4146
Krueger JG, Garber EA, Goldberg AR, Hanafusa H (1982) Changes in aminoterminal sequences of pp60^{src} lead to decreased membrane association and decreased in vivo tumorigenicity. Cell 28:889–896
Krzyzek RA, Mitchell RL, Lau AF, Faras AJ (1980) Association of pp60^{src} and *src* protein kinase activity with the plasma membrane of nonpermissive and permissive avian sarcoma virus-infected cells. J Virol 36:805–815
Lau AF, Krzyzek RA, Brugge JS, Erikson RL, Schollmeyer J, Faras AJ (1979) Morphological revertants of an avian sarcoma virus-transformed mammalian cell line exhibit tumorigenicity and contain pp60^{src}. Proc Natl Acad Sci USA 76:3904–3908
Lau AF, Krzyzek RA, Faras AJ (1981) Loss of tumorigenicity correlates with a reduction in pp60^{src} kinase activity in a revertant subclone of avian sarcoma virus-infected field vole cells. Cell 23:815–823
Lee JS, Varmus HE, Bishop JM (1979) Virus-specific messenger RNAs in permissive cells infected by avian sarcoma virus. J Biol Chem 254:8015–8022

Levinson AD, Oppermann H, Levintow L, Varmus HE, Bishop JM (1978) Evidence that the transforming gene of avian sarcoma virus encodes a protein kinase associated with a phosphoprotein. Cell 15:561–572

Levinson AD, Oppermann H, Varmus HE, Bishop JM (1980) The purified product of the transforming gene of avian sarcoma virus phosphorylates tyrosine. J Biol Chem 255:11973–11980

Levinson AD, Courtneidge SA, Bishop JM (1981) Structural and functional domains of the Rous sarcoma virus transforming protein (pp60^{src}). Proc Natl Acad Sci USA 78:1624–1628

Lipsich LA, Cutt JR, Brugge JS (1982) Association of the transforming proteins of Rous, Fujinami, and Y73 avian sarcoma viruses with the same two cellular proteins. Mol Cell Biol 2:875–880

Maness P, Engeser H, Greenberg ME, O'Farrell M, Gall WE, Edelman GM (1979) Characterization of the protein kinase activity of avian sarcoma virus *src* gene product. Proc Natl Acad Sci USA 76:5028–5032

Martin GS (1970) Rous sarcoma virus: a function required for the maintenance of the transformed state. Nature 227:1021–1023

Martinez R, Nakamura KD, Weber MJ (1982) Identification of phosphotyrosine-containing proteins in untransformed and Rous sarcoma virus-transformed chicken embryo fibroblasts. Mol Cell Biol 2:653–665

McGrath JP, Levinson AD (1982) Bacterial expression of an enzymatically active protein encoded by RSV *src* gene. Nature 295:423–425

Nakamura KD, Weber MJ (1982) Phosphorylation of a 36,000 M_r cellular protein in cells infected with partial transformation mutants of Rous sarcoma virus. Mol Cell Biol 2:147–153

Nakamura KD, Martinez R, Weber MJ (1983) Tyrosine phosphorylation of specific proteins after mitogen stimulation of chicken embryo fibroblasts. Mol Cell Biol 3:380–390

Neville DM (1975) Isolation of cell surface membrane fractions from mammalian cells and organs. In: Korn ED (ed) Methods in membrane biology, vol 3. Plenum, New York, pp 1–49

Nigg EA, Sefton BM, Hunter T, Walter G, Singer SJ (1982a) Immunofluorescent localization of the transforming protein of Rous sarcoma virus with antibodies against a synthetic *src* peptide. Proc Natl Acad Sci USA 79:5322–5326

Nigg EA, Walter G, Singer SJ (1982b) On the nature of crossreactions observed with antibodies directed to defined epitopes. Proc Natl Acad Sci USA 79:5939–5943

O'Farrell PH (1975) High resolution two-dimensional electrophoresis of proteins. J Biol Chem 250:4007–4021

O'Farrell PZ, Goodman HM, O'Farrell PH (1977) High resolution two-dimensional electrophoresis of basic as well as acidic proteins. Cell 12:1133–1142

Oppermann H, Levinson AD, Levintow L, Varmus HE, Bishop JM, Kawai S (1981a) Two cellular proteins that immunoprecipitate with the transforming protein of Rous sarcoma virus. Virology 113:736–751

Oppermann H, Levinson W, Bishop JM (1981b) A cellular protein that associates with the transforming protein of Rous sarcoma virus is also a heat-shock protein. Proc Natl Acad Sci USA 78:1067–1071

Osborn M, Weber K (1977) The detergent-resistant cytoskeleton of tissue culture cells includes the nucleus and the microfilament bundles. Exp Cell Res 106:339–349

Owada M-K, Donner P, Scott A, Moelling K (1981) Isolation of an avian sarcoma virus-specific protein kinase from virus particles. Virology 110:333–343

Pastan IH, Willingham M, de Crombrugghe B, Gottesman MM (1982) Aging and cancer: cyclic 3′,5′-adenosine monophosphate and altered gene activity. Natl Cancer Inst Monogr 60:7–15

Purchio AF, Erikson E, Erikson RL (1977) Translation of 35S and of subgenomic regions of avian sarcoma virus RNA. Proc Natl Acad Sci USA 74:4661–4665

Purchio AF, Jananovich S, Erikson RL (1980) Sites of synthesis of viral proteins in avian sarcoma virus-infected chicken cells. J Virol 35:629–636

Quinn PJ (1976) The molecular biology of cell membranes. University Park Press, Baltimore

Radke K, Martin GS (1979) Transformation by Rous sarcoma virus: effects of the *src* gene expression on the synthesis and phosphorylation of cellular polypeptides. Proc Natl Acad Sci USA 76:5212–5216

Radke K, Martin GS (1980) Transformation by Rous sarcoma virus: effects of the *src* gene expression on the synthesis and phosphorylation of cellular polypeptides. Cold Spring Harbor Symp Quant Biol 44:975–982

Radke K, Gilmore T, Martin GS (1980) Transformation by Rous sarcoma virus: a cellular protein substrate for transformation-specific protein phosphorylation contains phosphotyrosine. Cell 21:821–828

Revel JP, Wolken K (1973) Electron microscope investigations of the underside of cells in culture. Exp Cell Res 78:1–14

Richert ND, Blithe DB, Pastan IH (1982) Properties of the *src* kinase purified from Rous sarcoma virus-induced rat tumors. J Biol Chem 257:7143–7150

Robinson GB (1975) The isolation and composition of membranes. In: Parsons DS (ed) Biological membranes. Clarendon, Oxford, pp 8–54

Rohrschneider LR (1979) Immunofluorescence on avian sarcoma virus-transformed cells: localization of the *src* gene product. Cell 16:11–24

Rohrschneider LR (1980) Adhesion plaques of Rous sarcoma virus-transformed cells contain the *src* gene product. Proc Natl Acad Sci USA 77:3514–3518

Rohrschneider LR, Rosok MJ (1983) Transformation parameters and pp60^{src} localization in cells infected with partial transformation mutants of Rous sarcoma virus. Mol Cell Biol 3:731–746

Rohrschneider LR, Rosok M, Shriver K (1982) Mechanism of transformation by Rous sarcoma virus: events within adhesion plaques. Cold Spring Harbor Symp Quant Biol 46:953–965

Rosok MJ, Rohrschneider LR (1983) Increased phosphorylation of vinculin on tyrosine does not occur during the release of stress fibers before mitosis in normal cells. Mol Cell Biol 3:475–479

Rous P (1911) A sarcoma of the fowl transmissible by an agent separable from the tumor cells. J Exp Med 13:397–411

Rubsamen H, Friis RR, Bauer H (1979) *src* Gene product from different strains of avian sarcoma virus: kinetics and possible mechanism of heat inactivation of protein kinase activity from cells infected by transformation-defective, temperature-sensitive mutant and wildtype virus. Proc Natl Acad Sci USA 76:967–971

Rubsamen H, Saltenberger K, Friis RR, Eigenbrodt E (1982) Cytosolic malic dehydrogenase activity is associated with a putative substrate for the transforming gene product of Rous sarcoma virus. Proc Natl Acad Sci USA 79:228–232

Schaffhausen BS, Dorai H, Arakere G, Benjamin TL (1982) Polyoma virus middle T antigen: relationship to cell membranes and apparent lack of ATP-binding activity. Mol Cell Biol 2:1187–1198

Schlesinger MJ, Magee AI, Schmidt MFG (1980) Fatty acid acylation of proteins in cultured cells. J Biol Chem 255:10021–10024

Schliwa M, van Blerkom J, Porter KR (1981) Stabilization of the cytoplasmic ground substance in detergent opened cells and a structural and biochemical analysis of its composition. Proc Natl Acad Sci USA 78:4329–4333

Schmidt MFG (1982) Acylation of viral spike glycoproteins: a feature of enveloped RNA viruses. Virology 116:327–338

Schmidt MFG, Schlesinger MJ (1979) Fatty acid binding to vesicular stomatitis virus glycoprotein: a new type of post-translational modification of viral glycoproteins. Cell 17:813–819

Schmidt MFG, Schlesinger MJ (1980) Relation of fatty acid attachment to the translation and maturation of vesicular stomatitis and Sindbis virus membrane glycoproteins. J Biol Chem 255:3334–3339

Schmidt MFG, Bracha M, Schlesinger MJ (1979) Evidence for covalent attachment of fatty acids to Sindbis virus glycoproteins. Proc Natl Acad Sci USA 76:1687–1691

Schultz AM, Oroszlan S (1983) In vivo modification of retroviral *gag* gene-encoded polyproteins by myristic acid. J Virol 46:355–361

Sefton BM, Walter G (1982) Antiserum specific for the carboxy terminus of the transforming protein of Rous sarcoma virus. J Virol 44:467–474

Sefton BM, Beemon K, Hunter T (1978) Comparison of the expression of the *src* gene of the Rous sarcoma virus in vitro and in vivo. J Virol 28:957–971
Sefton BM, Hunter T, Beemon K (1979) Product of in vitro translation of the Rous sarcoma virus *src* gene has protein kinase activity. J Virol 30:311–318
Sefton BM, Hunter T, Beemon K (1980a) Temperature-sensitive transformation by Rous sarcoma virus and temperature-sensitive protein kinase activity. J Virol 33:220–229
Sefton BM, Hunter T, Beemon K, Eckhart W (1980b) Evidence that the phosphorylation of tyrosine is essential for cellular transformation by Rous sarcoma virus. Cell 20:807–816
Sefton BM, Hunter T, Ball EH, Singer SJ (1981) Vinculin: a cytoskeletal target of the transforming protein of Rous sarcoma virus. Cell 24:165–174
Sefton BM, Hunter T, Nigg EA, Singer SJ, Walter G (1982a) Cytoskeletal targets for viral transforming proteins with tyrosine protein kinase activity. Cold Spring Harbor Symp Quant Biol 46:939–951
Sefton BM, Trowbridge IS, Cooper JA, Scolnick EM (1982b) The transforming proteins of Rous sarcoma virus, Harvey sarcoma virus, and Abelson virus contain tightly-bound lipid. Cell 31:465–474
Shriver K, Rohrschneider LR (1981a) Organization of pp60^{src} and selected cytoskeletal proteins within adhesion plaques and junctions of Rous sarcoma virus-transformed rat cells. J Cell Biol 89:525–535
Shriver K, Rohrschneider LR (1981b) Spatial and enzymatic interaction of pp60^{src} with cytoskeletal proteins in isolated adhesion plaques and junctions from RSV-transformed NRK cells. Cold Spring Harbor Conf Cell Prolif 8:1247–1262
Singer I (1979) The fibronexus: a transmembrane association of fibronectin-containing fibers and bundles of 5 nm microfilaments in hamster and human fibroblasts. Cell 16:675–685
Singer I, Paradiso PR (1981) A transmembrane relationship between fibronectin and vinculin (130 kd protein): serum modulation in normal and transformed hamster fibroblasts. Cell 24:481–492
Singer SJ (1974) The molecular organization of membranes. Ann Rev Biochem 43:805–833
Smuckler EA, Koplitz M, Smuckler DE (1976) Isolation of animal cell nuclei. In: Birnie GD (ed) Subnuclear components; preparation and fractionation. Butterworth, Boston, pp 1–58
Takeya T, Hanafusa H (1982) DNA sequence of the viral and cellular *src* gene of chickens. II. Comparison of the *src* genes of two strains of ASV and of the cellular homolog. J Virol 44:12–18
Takeya T, Feldman RA, Hanafusa H (1982) DNA sequence of the viral and cellular *src* gene of chickens. I. Complete nucleotide sequence of an *Eco*RI fragment of recovered avian sarcoma virus which codes for gp37 and pp60^{src}. J Virol 44:1–11
Trotter JA, Foerder BA, Keller JM (1978) Intracellular fibres in cultured cells: analysis by scanning and transmission microscopy and by SDS-polyacrylamide gel electrophoresis. J Cell Sci 31:369–392
Wang E, Yin HL, Krueger JG, Caliguiri LA, Tamm I (to be published) Unphosphorylated gelsolin is localized in regions of cell substratum contact or attachment in Rous sarcoma virus-transformed rat cells. J Cell Biol
Warren L, Glick MC (1971) The isolation of surface membranes of animal cells: a survey. In: Manson LA (ed) Biomembranes, vol 1. Plenum, New York, pp 257–288
Warren L, Glick MC, Nass MK (1966) Membranes of animal cells. I. Methods of isolation of the surface membrane. J Cell Physiol 68:269–288
Webster RE, Henderson D, Osborn M, Weber K (1978) Three-dimensional electron microscopical visualization of the cytoskeleton of animal cells: immunoferritin identification of actin- and tubulin-containing structures. Proc Natl Acad Sci USA 75:5511–5515
Wehland J, Osborn M, Weber K (1979) Cell-to-substratum contacts in living cells: a direct correlation between interference-reflection and indirect-immunofluorescence microscopy using antibodies against actin and alpha-actinin. J Cell Sci 37:257–273
Weiss R, Teich N, Varmus H, Coffin J (eds) (1982) RNA tumor viruses. Molecular biology of tumor viruses, 2nd ed. Cold Spring Harbor Laboratory, Cold Spring Harbor, New York

Willingham MC, Yamada SS, Pastan I (1978) Ultrastructural antibody localization of alpha-$_2$-macroglobulin in membrane-limited vesicles in cultured cells. Proc Natl Acad Sci USA 75:4359–4363
Willingham MC, Jay G, Pastan I (1979) Localization of ASV *src* gene product to the plasma membrane of transformed cells by electron microscopic immunocytochemistry. Cell 18:125–134
Willingham MC, Pastan I, Shih TY, Scolnick EM (1980) Localization of the *src* gene product of the Harvey strain of MSV to plasma membrane of transformed cells by electron microscopic immunocytochemistry. Cell 19:1005–1014
Witt DP, Gordon JA (1980) Specific dephosphorylation of membrane proteins in Rous sarcoma virus-transformed chick embryo fibroblasts. Nature 287:241–244
Yu J, Fischman DA, Steck TL (1973) Selective solubilization of proteins and phospholipids from red blood cell membranes by nonionic detergents. J Supramol Struct 1:233–248

Regulation of Cell Growth and Transformation by Tyrosine-Specific Protein Kinases: The Search for Important Cellular Substrate Proteins

JONATHAN A. COOPER and TONY HUNTER

1 Background 125
2 Proteins Phosphorylated at Tyrosine in Living Cells 131
2.1 Tyrosine Protein Kinases 131
2.2 The 50-K Protein 135
2.3 The 36-K Protein 136
2.4 Vinculin 144
2.5 Glycolytic Enzymes 146
2.6 Other Substrates 150
3 One Critical Substrate or Many? 150
References 153

1 Background

It is now well established that post-translational modification of vertebrate cell proteins can occur by phosphorylation at serine, threonine, and tyrosine. Many serine- and threonine-specific protein kinases and their substrates have been extensively characterized, and in several instances protein phosphorylation has been shown to play an important role in the regulation of cell metabolism through alteration of the properties of specific enzymes. Historically, evidence that modulation of protein function led to a significant change in metabolism often predated the discovery that the protein in question was a phosphoprotein and the isolation of the regulatory protein kinase. For example, increased glycogenolysis in liver slices, under conditions of adrenaline or glucagon treatment, was correlated with activation of the enzyme glycogen phosphorylase. The active form of glycogen phosphorylase was subsequently found to be a phosphoprotein, and the protein kinase responsible, phosphorylase kinase, was characterized (KREBS and FISCHER 1956). The regulation of phosphorylase kinase itself by phosphorylation has since been elucidated in great detail (COHEN 1978).

In contrast, tyrosine protein kinases were discovered before their substrates, and modulation of a function of any individual protein substrate by a tyrosine protein kinase has yet to be described. However, the gross distal effects of tyrosine protein kinases can be highly conspicuous at the

Molecular Biology and Virology Laboratory, The Salk Institute, P.O. Box 85800, San Diego, CA 92138, USA

Current Topics in Microbiology and Immunology, Vol. 107

cellular level, since tyrosine protein kinase activity is a property of many retroviral transforming proteins and of some cellular receptors for polypeptide regulators. The activity of a virally-coded tyrosine protein kinase accompanies, and may well cause, malignant transformation. The activation of a cellular growth factor receptor results in the arousal of quiescent cell cultures. Given such dramatic effects of tyrosine phosphorylation, we anticipate that examples of regulation of substrate protein function by tyrosine protein kinases will be soon be established.

Some properties of known tyrosine protein kinases are summarized in Table 1 and are discussed in detail elsewhere (Hunter and Sefton 1982, Bishop and Varmus 1982). A plethora of different tyrosine protein kinases have been described in the few years since the first examples were discovered. This is partly due to the incorporation of a number of tyrosine protein kinase genes into retroviruses (Bishop and Varmus 1982)[1]. These viruses were selected in the wild and the laboratory for their ability to induce malignant transformation. Analysis of their transforming genes and products was facilitated by the ready supply of nucleic acid probes and antisera. Serendipitously, the transforming proteins which possess tyrosine protein kinase activity appear to retain that ability even when bound to antibody in immunoprecipitates, so both their assay and their separation from the bulk of protein kinases of the host cell are straightforward. Of the 16 genetically distinguishable groups of retroviruses presently recognized (Bishop and Varmus 1982), seven appear to encode tyrosine protein kinases (Table 1). So far there is no clear example of a retroviral transforming protein with serine or threonine protein kinase activity, although it is possible that such kinases are not as amenable to assay in immunoprecipitates as tyrosine protein kinases.

Inappropriate tyrosine protein kinase activity may be especially disruptive to cellular growth control mechanisms. This would be likely if phosphorylation of tyrosine in certain key regulatory proteins is a factor in normal mechanisms of cell growth control. The detection of tyrosine protein kinase activities tightly associated with three cellular proteins which are all cell surface receptors for polypeptide growth factors (Table 1) suggests just such a link between tyrosine phosphorylation and cell growth. Moreover, many viral transforming proteins, like growth factor receptors, are associated with

1 The question of whether all these transforming proteins (Table 1) are themselves protein kinases or are complexed with protein kinases is still moot. One example of a transforming protein which may not itself be a tyrosine protein kinase but is phosphorylated in an immunoprecipitate, is the medium-sized tumor antigen of the DNA virus, polyoma virus (Eckhart et al. 1979). In contrast, the transforming protein of RSV, pp60$^{v\text{-}src}$, is well established as a tyrosine protein kinase since this protein has tyrosine protein kinase activity even when made in bacterial cells containing suitable recombinant plasmids (Gilmer and Erikson 1981; McGrath and Levinson 1982). The case is not watertight for all the other viral protein kinases listed in Table 1, but predicted amino acid sequences suggest that at least a portion of each protein is highly related to pp60$^{v\text{-}src}$ (Kitamura et al. 1982; Shibuya and Hanafusa 1982; Hampe et al. 1982), so they are most likely authentic tyrosine protein kinases. In some cases there is evidence that the protein products of the homologous cellular genes (c-*onc* genes) also have tyrosine protein kinase activity (Table 1)

membranes. A working hypothesis is that some retroviral transforming proteins have the same enzymatic activity and overlapping protein specificity as some growth factor receptors but are not subject to the same regulation. Retroviruses encoding tyrosine protein kinases would subvert the normal cellular response to growth factors, and a condition of uncontrolled cell proliferation would result.

The reasons why most tyrosine protein kinases presently known should be "growth regulatory", rather than controlling, say, intermediary metabolism, are partly historical and partly evolutionary. Historically, once phosphotyrosine was identified as the product of $pp60^{v\text{-}src}$ (HUNTER and SEFTON 1980) and of an EGF-stimulated activity in EGF receptor-rich cell membranes (USHIRO and COHEN 1980), it became *de rigeur* to perform phosphoamino acid analyses to check for phosphotyrosine in the proteins phosphorylated by other viral transforming proteins and in proteins phosphorylated in cell membranes incubated with other growth factors. From an evolutionary standpoint, many retroviral transforming genes, although judged to be unrelated by nucleic acid hybridization or by peptide mapping of their protein products, are now proving to have extensive homologies at the amino acid sequence level and may thus have a common origin (KITAMURA et al. 1982; SHIBUYA and HANAFUSA 1982; HAMPE et al. 1982). Too little is known about the structures of growth factor receptors to say whether they share a similarly related protein kinase domain. Although this seems probable one should not rule out the possibility of another family of tyrosine protein kinases, unrelated to the others in function or ancestry. Whatever the significance of the amino acid specificity of tyrosine protein kinases, it has facilitated the identification of their substrates, since the phosphorylation of tyrosine is a rare protein modification (SEFTON et al. 1980b).

The scarcity of phosphotyrosine in proteins means that it is difficult to detect proteins modified in this way. The amounts of phosphoserine and phosphothreonine in transformed cell proteins outweigh the amount of phosphotyrosine by 500-fold and 50-fold respectively (SEFTON et al. 1980b). However, sensitive techniques for the detection of phosphotyrosine in proteins have been developed and are described elsewhere (COOPER et al. 1983c). Phosphotyrosine has been identified in phosphoproteins isolated from living cells as well as in the products of cell-free reactions. (We will use the terms "in vivo" and "in vitro" to describe these respective phosphorylation conditions.) The extent of protein phosphorylation at tyrosine decreases rapidly in RSV-transformed cells when the kinase activity of a thermolabile $pp60^{v\text{-}src}$ is inactivated by increased temperature (SEFTON et al. 1980b). This suggests that the extent of phosphorylation of individual tyrosine protein kinase substrates in the cell is controlled not only by specific kinases but also by specific phosphatases (FOULKES et al. 1981, 1983; BRAUTIGAN et al. 1981). The specificities of individual tyrosine protein kinases and phosphatases may be partly due to intrinsic properties which might be expected to operate in vitro as well as in vivo. In addition the activities of these enzymes may be affected by other proteins or by the subcellular locations of kinase, substrate, and phosphatase.

Table 1. Proteins associated with tyrosine protein kinase activity

Protein[a]	Gene found in	Gene name	Reference[b]	Cellular protein[c]		
				Size	Tyrosine protein kinase?	Reference[d]
$pp60^{v\text{-}src}$	Rous sarcoma virus (RSV)	*src*	1–14	60 K	+	10,11, 37–40
$P90^{gag\text{-}yes}$ $P80^{gag\text{-}yes}$	Y73 avian sarcoma virus Esh sarcoma virus (ESV)	*yes*	15, 16 17	?	?	–
$P140^{gag\text{-}fps}$ $P105^{gag\text{-}fps}$ $P150^{gag\text{-}fps}$	Fujinami sarcoma virus (FSV) PRC II sarcoma virus UR-1 sarcoma virus	*fps*[e]	18–21 20, 22–24 25	98 K	+	41
$P85^{gag\text{-}fes}$ $P95^{gag\text{-}fes}$	Snyder-Theilen feline sarcoma virus (STFeSV) Gardner-Arnstein feline sarcoma virus	*fes*[e]	20, 26–29 27, 28, 30	92 K	?	26
$P68^{gag\text{-}ros}$	UR-2 sarcoma virus[f]	*ros*	31	?	?	–
$P120^{gag\text{-}abl}$	Abelson murine leukemia virus (AMuLV)[g]	*abl*	28, 32–35	150 K	–	42, 43
$P72^{gag\text{-}fgr}$	Gardner-Rasheed feline sarcoma virus[h]	*fgr*	36	?	?	–
EGF receptor	(cellular)	?	–	170 K	EGF-stimulated	44–48
PDGF receptor	(cellular)	?		175 K	PDGF-stimulated	49–51
Insulin receptor	(cellular)	?		130 K, 95 K[i]	Insulin-stimulated	52–54
?	LSTRA cells (cellular or viral?)[j]	?	–	?	+	55–56

[a] Viral transforming proteins are designated by their size, in kilodaltons, and genetic origin (most are encoded by fused *gag-onc* genes in the viral genome); cellular proteins are designated by their presumed function

[b] References to the association of protein kinase activity with the viral proteins, and to the identification of the target amino acid as tyrosine. In some cases, genetic or biochemical evidence that the tyrosine protein kinase activity is integral to the transforming protein is also cited

1 COLLETT and ERIKSON (1978)
2 LEVINSON et al. (1978)
3 SEFTON et al. (1979)
4 MANESS et al. (1979)
5 RICHERT et al. (1979)
6 RUBSAMEN et al. (1979)
7 ERIKSON et al. (1979)
8 SEFTON et al. (1980a)
9 HUNTER and SEFTON (1980)
10 COLLETT et al. (1980)
11 LEVINSON et al. (1980)
12 GILMER and ERIKSON (1981)
13 MCGRATH and LEVINSON (1982)
14 RICHERT et al. (1982)
15 KAWAI et al. (1980)
16 GHYSDAEL et al. (1981a)
17 GHYSDAEL et al. (1981b)
18 FELDMAN et al. (1980)
19 PAWSON et al. (1980)
20 BEEMON (1981)
21 HANAFUSA et al. (1981)
22 NEIL et al. (1981b)
23 HIRANO and VOGT (1981)
24 ADKINS et al. (1982a)
25 WANG et al. (1981)
26 BARBACID et al. (1980)
27 REYNOLDS et al. (1981b)
28 VAN DE VEN et al. (1980)
29 BARBACID et al. (1981b)
30 REYNOLDS et al. (1980b)
31 FELDMAN et al. (1982)
32 WITTE et al. (1980a)
33 WITTE et al. (1980b)
34 REYNOLDS et al. (1980a)
35 WANG et al. (1982)
36 SEFTON BM, RASHEED S, personal communication

[c] Size of product of homologous cellular *onc* gene in the case of the viral transforming proteins

[d] References to the association of protein kinase activity with the cellular proteins, and to the identification of the target amino acid as tyrosine. In some cases, biochemical evidence that the tyrosine protein kinase is integral to the protein is also cited

37 COLLETT et al. (1979)
38 OPPERMAN et al. (1979)
39 ROHRSCHNEIDER et al. (1979)
40 HUNTER and SEFTON (1980)
41 MATHEY-PREVOT et al. (1982)
42 WITTE et al. (1979)
43 PONTICELLI et al. (1982)
44 CARPENTER et al. (1979)
45 COHEN et al. (1980)
46 USHIRO and COHEN (1980)
47 BUHROW et al. (1982)
48 COHEN et al. (1982)
49 EK et al. (1982)
50 NISHIMURA et al. (1982)
51 EK and HELDIN (1982)
52 KASUGA et al. (1981)
53 AVRUCH et al. (1982)
54 PETRUZZELLI et al. (1982)
55 CASNELLIE et al. (1982a)
56 CASNELLIE et al. (1982b)

[e] *fes* and *fps* are closely related genes (SHIBUYA et al. 1970) encoding similar proteins (BEEMON 1981; BARBACID et al. 1981a) which probably arose from the homologous cellular genes in two species. Presumably *fps* was recovered by infection of feline cells, whereas *fps* was recovered from avian cells

[f] The transforming protein of UR-2 sarcoma virus is included here since, like the other viral tyrosine protein kinases, both the transforming protein and immunoglobulin become phosphorylated at tyrosine when ATP is added to an immunoprecipitate, and the transforming protein is itself phosphorylated in transformed cells (FELDMAN et al. 1982). However, it has proved difficult to demonstrate elevated phosphotyrosine levels either in total proteins from UR-2 infected chicken embryo cells or in clones of rat cells transformed by the virus (PATSCHINSKY T, SEFTON BM, personal communication), and there is minimal phosphorylation of individual target proteins as detected by two-dimensional gel electrophoresis (JAC, PATSCHINSKY T, SEFTON BM, unpublished data). Therefore, the tyrosine protein kinase associated with $P68^{gag\text{-}ros}$ seems to be of limited protein specificity

[g] Other isolates of Abelson murine leukemia virus encode transforming proteins of various sizes, from 160 K to 90 K (see ROSENBERG 1982)

[h] Incompletely characterized virus isolated by S. RASHEED (RASHEED et al. 1982). Its relationship to the other retroviruses listed here is not completely known

[i] Probably an a_2b_2 structure. The smaller subunit is phosphorylated at tyrosine when insulin is bound

[j] LSTRA cells contain an active tyrosine protein kinase of unknown provenance. The primary endogenous substrate is a 53-58-K protein which may be the kinase itself. This protein appears to be phosphorylated at tyrosine in intact LSTRA cells (SEFTON BM, COOPER JA, unpublished data)

Table 2. Proteins phosphorylated at tyrosine in vivo[a]

Viral transforming proteins	Cellular proteins
$pp60^{v\text{-}src}$	$pp60^{c\text{-}src}$
$P90^{gag\text{-}yes}$	EGF receptor
$P80^{gag\text{-}yes}$	PDGF receptor?
$P140^{gag\text{-}fps}$	Insulin receptor
$P105^{gag\text{-}fps}$ (and proteins of viruses related to PRCII)	pp50
	pp36
$P150^{gag\text{-}fps}$	Vinculin
$P85^{gag\text{-}fes}$	Enolase
$P95^{gag\text{-}fes}$	Phosphoglycerate mutase
$P68^{gag\text{-}ros}$	Lactate dehydrogenase
$P120^{gag\text{-}abl}$ (and variant AMuLV proteins)	81-K protein
	42-K to 45-K proteins
$P72^{gag\text{-}fgr}$	LSTRA cell protein

[a] These phosphoproteins are discussed in more detail in Sect. 2

In practice, tyrosine protein kinases can phosphorylate many proteins in vitro which do not appear to be substrates in vivo. For example, there is no evidence that histones are phosphorylated at tyrosine under any conditions in vivo, yet they are good substrates for the EGF receptor-associated tyrosine protein kinase (COHEN et al. 1980). Authentic in vivo substrates can also be phosphorylated artifactually in vitro. For example, the EGF receptor and AMuLV $P120^{gag\text{-}abl}$ appear to be phosphorylated at only one or two tyrosine residues in vivo, but can "autophosphorylate" at many tyrosines when incubated in vitro with ATP (HUNTER and COOPER 1981; HUNTER and COOPER 1983a; WITTE et al. 1980a; SEFTON et al. 1981b). These discrepancies between in vivo and in vitro observations could, of course, be due to insensitivity of the in vivo measurements or to differential protein phosphatase activities under the different conditions. On the other hand, it seems more likely that the protein specificity (but not tyrosine specificity) of tyrosine protein kinases is lost when they are extracted from the cell. Fidelity may normally be ensured by combined effects of compartmentalization and by loose association of other molecules with kinase or substrate. Considerable caution is therefore advised in interpreting tyrosine phosphorylations detected in broken cell systems or partially purified systems, unless the proteins phosphorylated are also shown to be phosphorylated at the same site(s) in intact cells. In several notable instances, however, authentic in vivo substrates *are* phosphorylated faithfully in vitro, and in these cases in vitro studies complement the in vivo studies.

So far (January 1983), phosphotyrosine has been found in 11 viral proteins and 13 cellular proteins, after their isolation from cells (Table 2). For most such cellular proteins, the level of phosphotyrosine is increased dramatically when a particular tyrosine protein kinase is present, or activated, in the cell. In cells having high levels of tyrosine protein kinase activity, phosphotyrosine constitutes only about 0.2% of acid-stable protein phosphate (SEFTON et al. 1980b) but is found in proteins of all sizes (BEEMON et al. 1982; MARTINEZ et al. 1982). More phosphotyrosine-containing proteins may yet be discovered. However, there are many proteins which do not contain phosphotyrosine at detectable levels. Examples include actin, α-actinin, and myosin (SEFTON et al. 1981a), and a large number of phosphoproteins identified by two-dimensional gel electrophoresis (COOPER and HUNTER 1981a). Many of the proteins which *do* contain phosphotyrosine are known only by their physical properties but some have ascribed functions. Both categories are discussed in detail below.

Intensive characterization of substrate proteins for tyrosine protein kinases is important because of the expectation that they may be involved in the control of cell growth in normal and malignant cells. Such a connection has yet to be established for any substrate. Ultimately, in vivo studies, correlating phosphorylation state in relation to cell physiology, should be complemented by in vitro studies, exploring the enzymology of the kinase/substrate and phosphorylated substrate/phosphatase interactions, and the effect of the phosphorylation on substrate function. Ideally, a complete description of a biologically significant phosphorylation should address the following questions:

1. Under what conditions, at what sites, and to what extent is the substrate phosphorylated on tyrosine in vivo?
2. Can the substrate be phosphorylated in vitro by the purified tyrosine protein kinase in question, at the appropriate site?
3. What is the function of the substrate in cell physiology?
4. Is this function modulated by phosphorylation of the substrate?
5. Does the altered function contribute toward a change in cell physiology?
6. How is the extent of phosphorylation controlled in vivo by kinases and phosphatases?
7. Is phosphorylation or other modification of additional amino acids in the same protein conditional on the tyrosine phosphorylation, or is the converse true? Do these second site modifications affect function?

The following discussion of individual tyrosine protein kinase substrates is intended to show the limits of our knowledge on the answers to these questions rather than to supply a comprehensive review of the literature.

2 Proteins Phosphorylated at Tyrosine in Living Cells

2.1 Tyrosine Protein Kinases

All tyrosine protein kinases characterized to date are themselves substrates for tyrosine protein kinase(s). Historically, $pp60^{v\text{-}src}$ was the first protein shown to contain phosphotyrosine in vivo (HUNTER and SEFTON 1980).

About 30% of pp60$^{v\text{-}src}$ molecules are phosphorylated on tyrosine, at a single major site (SEFTON et al. 1982). pp60$^{v\text{-}src}$ is also phosphorylated at serine, as are most of the other tyrosine protein kinases. Y73 virus P90$^{gag\text{-}yes}$, AMuLV P120$^{gag\text{-}abl}$, and the other viral transforming proteins listed in Table 1 all contain one or more sites of tyrosine phosphorylation when isolated from transformed cells (NEIL et al. 1981 a, c; PATSCHINSKY and SEFTON 1981; PATSCHINSKY et al. 1982; REYNOLDS et al. 1982a, b; SMART et al. 1981; SEFTON et al. 1981 b; WITTE et al. 1981; and references in Table 1). The EGF receptor and insulin receptor contain much more phosphotyrosine when isolated from cells treated with the respective growth factor than they contain when isolated from untreated cells (HUNTER and COOPER 1981, 1983a; KASUGA et al. 1981). In most cases, the principal in vivo tyrosine phosphorylation sites in these viral and cellular tyrosine protein kinases can also be phosphorylated in vitro, when the partially purified kinase is incubated with ATP. In the case of the EGF receptor and AMuLV P120$^{gag\text{-}abl}$, additional tyrosines are phosphorylated in vitro (HUNTER and COOPER 1983a; SEFTON et al. 1981 b; WITTE et al. 1981). It is possible that these "bogus" sites are normally masked by loosely associated proteins or membrane lipids in vivo. Phosphorylation of AMuLV P120$^{gag\text{-}abl}$ is apparently limited to the authentic sites if the in vitro reaction is short and at low temperature (REYNOLDS et al. 1982b).

It is currently unclear (a) whether these tyrosine phosphorylations induce functional changes in the protein kinases, and (b) whether the phosphorylations are genuine autophosphorylations or are catalyzed by associated tyrosine protein kinases of unknown provenance.

Is the phosphorylation of tyrosine protein kinases on tyrosine of physiological significance? On the one hand, a requirement for phosphorylation for maximal activity of protein kinase activity is consistent with the following observations. The normal cell homologues of pp60$^{v\text{-}src}$ and AMuLV P120$^{gag\text{-}abl}$ have been isolated by immunoprecipitation. The former is both phosphorylated at tyrosine and competent in protein kinase assays employing exogenous or endogenous substrates (COLLETT et al. 1979; OPPERMANN et al. 1979). The AMuLV P120$^{gag\text{-}abl}$ homologue (p150$^{c\text{-}abl}$) is neither phosphorylated at tyrosine nor a tyrosine protein kinase, at least when assayed for "autophosphorylation" (PONTICELLI et al. 1982). It can be phosphorylated at tyrosine by added P120$^{gag\text{-}abl}$ protein kinase, but it is not known whether this leads to activation of a latent tyrosine protein kinase activity in p150$^{c\text{-}abl}$ (PONTICELLI et al. 1982). Phosphorylation of the EGF receptor at tyrosine accompanies increased tyrosine protein kinase activity in EGF-treated cells (HUNTER and COOPER 1981). On the other hand, the following observations argue against a role for tyrosine phosphorylation in kinase regulation. The FSV P140$^{gag\text{-}fps}$ cell homologue (p98$^{c\text{-}fps}$) is not detectably phosphorylated at tyrosine in vivo yet it does have tyrosine protein kinase activity (MATHEY-PREVOT et al. 1982). Mutant pp60$^{v\text{-}src}$ molecules in which the target tyrosine residue is replaced with a phenylalanine or is deleted together with some adjacent amino acids have normal in vitro kinase activity and can transform cells (SNYDER et al. 1983; CROSS and HANAFUSA 1983). Thus, phosphorylation of tyrosine in pp60$^{v\text{-}src}$ may not be absolutely necessary for its activity, inasmuch as this can be quantitated by these assays. One should be cautious, however, about extrapolating from these results to other viral or cellular tyrosine protein kinases.

The second question, whether tyrosine protein kinases phosphorylate themselves (either in *cis* or in *trans*), termed "autophosphorylation", or whether they are substrates for associated tyrosine protein kinases, is difficult to answer with protein kinases isolated from vertebrate cells, since their purity and integrity will always be uncertain. For example, pp60$^{v\text{-}src}$ preparations able to autophosphorylate may be contaminated with other kinases (ERIKSON et al. 1979), whilst impotent pp60$^{v\text{-}src}$ preparations may have been damaged during isolation (LEVINSON et al. 1980; BLITHE et al. 1982). Preparations of pp60$^{v\text{-}src}$ from transfected bacteria appear to be of low specific activity, both for autophosphorylation and for phosphorylation of added substrates (GILMER and ERIKSON 1981; MCGRATH and LEVINSON 1982). It is not clear whether the low autophosphorylation activity of such a preparation of pp60$^{v\text{-}src}$ is due to the absence of suitable tyrosine protein kinases in the bacterial cell or reflects its low specific activity overall. When large quantities of these under-phosphorylated bacterial products are available, they may be useful to assay for normal cell protein kinases able to phosphorylate the target tyrosine.

The above questions are important since the tyrosines which are phosphorylated in transforming proteins are the only in vivo phosphorylation sites for which sequences are known (Table 3). In the case of many of these protein kinases entire amino acid sequences are known from the gene sequences and the correct tyrosine in each sequence has been found by peptide mapping experiments and by partial sequencing of ^{32}P-labeled tryptic peptides (PATSCHINSKY and SEFTON 1981; SMART et al. 1981; PATSCHINSKY et al. 1982; NEIL et al. 1981c). So far, all the sequences embedding the target tyrosines are highly related and contain acidic residues to the amino terminal side. The sequence of the primary in vivo tyrosine phosphorylation site of the EGF receptor is not known in detail, but in this case the ability to autophosphorylate is retained through extensive purification, including nondenaturing polyacrylamide gel electrophoresis (COHEN et al. 1982). This site shares some common features with the sites in viral proteins (HUNTER and COOPER 1983a; Table 3). Synthetic peptides having the pp60$^{v\text{-}src}$ site sequence are substrates for viral and cellular tyrosine protein kinases in vitro, albeit with low affinity (CASNELLIE et al. 1982b; HUNTER 1982a; PIKE et al. 1982), suggesting that the local primary structure may indeed be an important specificity determinant. However, some synthetic peptides of apparently unrelated sequence are also substrates (WONG and GOLDBERG 1983). To what extent the similarity between phosphorylation sites in different transforming proteins results merely from their common evolutionary origins, and to what extent it defines the specificity of the relevant tyrosine protein kinase, is unknown.

Table 3. Sequences of tyrosine phosphorylation sites in tyrosine protein kinases[a]

Virus	Protein	Sequence																
		−9	−8	−7	−6	−5	−4	−3	−2	−1	0	1	2	3	4	5	6	7
Rous sarcoma virus	pp60$^{v\text{-}src}$			**arg**	leu	ile	*glu*	*asp*	asn	*glu*	**tyr**	thr	ala	**arg**	gln	gly	ala	**lys**
Y73 avian sarcoma virus	P90$^{gag\text{-}yes}$			**arg**	leu	ile	*glu*	*asp*	asn	*glu*	**tyr**	thr	ala	**arg**	gln	gly	ala	**lys**
Snyder-Theilen FeSV	P85$^{gag\text{-}fes}$			**arg**	*glu*	*glu*	ala	*asp*	gly	val	**tyr**	ala	ala	ser	gly	gly	leu	**arg**
Gardner-Arnstein FeSV	P110$^{gag\text{-}fes}$			**arg**	*glu*	ala	ala	*asp*	gly	ile	**tyr**	ala	ala	ser	gly	gly	leu	**arg**
Fujinami sarcoma virus	P140$^{gag\text{-}fps}$			**arg**	gln	*glu*	*glu*	*asp*	gly	val	**tyr**	ala	ser	thr	gly	gly	met	**lys**
(A431 cell)	EGF receptor	**lys)** **arg)**	x	x	x	x	*glu*	x	x	*glu*	**tyr**	x	x	x	x	x	x	x

[a] The sequences shown are of tryptic peptides which contain phosphotyrosine when tyrosine protein kinases are isolated from cells. In most cases the same peptides can be labeled in vitro by incubation with ATP. In each case the phosphorylated tryptic peptide was characterized by partial sequence analysis and by secondary cleavage with *S. aureus* V8 protease (specific for *glu*). The complete amino acid sequence of the peptide was then found in the predicted sequence of the viral transforming protein (CZERNILOFSKY et al. 1980; SCHWARTZ et al. 1982; TAKEYA et al. 1982; REDDY et al. 1983; KITAMURA et al. 1982; SHIBUYA and HANAFUSA 1982, and HAMPE et al. 1982). For pp60^{src} and Y73 P90 only, the assignation of amino acid sequence was confirmed by amino acid composition analysis and by chemical synthesis and comigration of the predicted peptide. References: SMART et al. (1981); PATSCHINSKY et al. (1982); NEIL et al. (1981c); HUNTER and COOPER (1983a); REYNOLDS et al. (1982a). Basic and acidic residues are in **bold** type, the acidic residues are *italicized*. The phosphorylated tyr is in position '0' in each case. 'x' signifies an unknown amino acid

2.2 The 50-K Protein

A 50-K protein (p50) was first identified as a specific component of immunoprecipitates of pp60$^{v\text{-}src}$, prepared from RSV-transformed chicken embryo cells (SEFTON et al. 1978), and was shown to be phosphorylated at a single tyrosine and a single serine (HUNTER and SEFTON 1980; OPPERMANN et al. 1981b). Phosphopeptide mapping and its constant size (in contrast to the polymorphism of pp60$^{v\text{-}src}$ of different RSV strains) suggested that pp50 was not a viral product (HUNTER and SEFTON 1980; SEFTON et al. 1978). Subsequently, the 50-K protein was shown to be complexed to a subpopulation of pp60$^{v\text{-}src}$ (BRUGGE et al. 1981) and was also associated with several other viral tyrosine protein kinases including Y73 P90$^{gag\text{-}yes}$, PRCII P105$^{gag\text{-}fps}$, and FSV P140$^{gag\text{-}fps}$ (LIPSICH et al. 1982; ADKINS et al. 1982a). Another phosphoprotein, of 80–90 K (pp90), which is phosphorylated at multiple serine sites but not at tyrosine, is complexed together with p50 and the transforming protein (SEFTON et al. 1978; OPPERMANN et al. 1981b; BRUGGE et al. 1981)[1]. pp50 and pp90 are not so readily detected in immunoprecipitates of pp60$^{v\text{-}src}$ prepared from RSV-transformed mammalian cells, and it is possible that this is the reason why they have not been found complexed with transforming proteins of mammalian retroviruses. It is not known whether these proteins associate with all tyrosine protein kinases, nor whether they are associated with any transforming proteins which are not tyrosine protein kinases.

The phosphorylation state of pp50 in normal cells could not be evaluated until pp50 was identified by virtue of some property other than its association with transforming proteins, which, obviously, are absent from normal cells (BRUGGE and DARROW 1982; GILMORE et al. 1982). Immunoprecipitates containing pp50 were analyzed by two-dimensional gel electrophoresis, and the coordinates of pp50 were found to correspond to a cellular phosphoprotein which contained phosphotyrosine and phosphoserine in transformed cells. A related phosphoprotein was identified in normal cells, which lacked phosphotyrosine (BRUGGE and DARROW 1982; GILMORE et al. 1982). Thus it seems likely that pp50 is a substrate for the tyrosine protein kinase with which it associates. However, pp50 complexed with the pp60$^{v\text{-}src}$ of mutants of RSV temperature sensitive for transformation is phosphorylated at tyrosine equally at the permissive and nonpermissive temperatures (BRUGGE and DARROW 1982; GILMORE et al. 1982). This contrasts with the phosphorylation of pp60$^{v\text{-}src}$ and of other cellular substrates. This paradox may be explained if the tight binding of pp50 to pp60$^{v\text{-}src}$ ensures its extensive phosphorylation even under conditions where pp60$^{v\text{-}src}$ has reduced activity toward other substrates.

Roles for pp50 and pp90 have been proposed. Both proteins are bound to newly synthesized, soluble pp60$^{v\text{-}src}$, but not to membrane-bound pp60$^{v\text{-}src}$

1 This is the same abundant 90-K protein which is induced in many vertebrate cells in response to stress, and whose levels are also affected by glucose availability (OPPERMANN et al. 1981a; LANKS et al. 1982). The significance of these properties in relation to the association with pp60$^{v\text{-}src}$ is a matter of speculation

(Brugge et al. 1983; Courtneidge and Bishop 1982). Thus they could mediate the transport of pp60$^{v\text{-}src}$ and other transforming proteins from the soluble ribosomes where they are synthesized to the plasma membrane. There are indications that pp50 of normal cells may exist mostly as a soluble complex with pp90 (Brugge JS, personal communication).

Although serine phosphorylated pp50 was found in uninfected cells, a nonphosphorylated form has not been detected. Thus there is no convenient way of recognizing the nonphosphorylated protein, if it exists, among other cell proteins of the same size. This has hindered efforts to purify the protein. Hence it is not known whether pp50 can be phosphorylated in vitro by purified pp60$^{v\text{-}src}$. It does not appear to be phosphorylated in vitro when immunoprecipitates of pp60$^{v\text{-}src}$ are incubated with ATP, but if the associated pp50 is already stoichiometrically phosphorylated before isolation of the complex from the cell then there may be no available sites.

2.3 The 36-K Protein

A 36-K basic phosphoprotein (pp36) was detected by Radke and Martin in RSV-transformed, but not normal, chicken embryo cells, and had the characteristics expected for a pp60$^{v\text{-}src}$ substrate (Radke and Martin 1979). It was phosphorylated rapidly when temperature-sensitive mutant-infected cells were shifted from nonpermissive to permissive temperature and was scarcely detectable in cells transformed by a virus with no known tyrosine protein kinase involvement, the acute avian leukemia virus, MC29 (Radke and Martin 1979). This was the first cellular substrate of a tyrosine protein kinase to be identified, although it was not until later that its phosphoamino acid composition was determined and phosphotyrosine and phosphoserine were found (Radke et al. 1980; Erikson and Erikson 1980; Cooper and Hunter 1981a). The facility of detection of this phosphoprotein has engendered studies by many investigators, who have assigned a plethora of molecular weights (34–39 K) to what is most likely the same phosphoprotein (Radke and Martin 1979; Erikson and Erikson 1980; Kobayashi and Kaji 1980; Cooper and Hunter 1981a; Cheng and Chen 1981; Nakamura and Weber 1982; Decker 1982; Courtneidge et al. 1983; Greenberg and Edelman 1983a). We will use the original molecular weight designation. It is possibly the most abundant phosphotyrosine-containing protein in an RSV-transformed cell (Martinez et al. 1982).

A cellular protein (p36) has been identified as the nonphosphorylated form of pp36 by several criteria. It has a similar size to pp36, an isoelectric point shifted from that of pp36 by the removal of one ionized phosphate group, and a similar peptide map to pp36, and both p36 and pp36 are recognized by antisera raised to the former (Radke et al. 1980; Erikson and Erikson 1980; Cooper and Hunter 1981b; Courtneidge et al. 1983; Cooper and Hunter 1983). p36 is a major protein of chick fibroblasts, comprising 0.2%–0.25% of total macromolecular methionine (Radke et al. 1980; Cooper and Hunter 1983). For a chicken fibroblast containing

5×10^{-10} g of protein, this calculates at about 1.5×10^7 molecules per cell. Because of its abundance and basic nature, p36 can be purified simply (ERIKSON and ERIKSON 1980), and specific antisera generated (ERIKSON et al. 1981b; COOPER and HUNTER 1983; COURTNEIDGE et al. 1983). A monoclonal antibody has been prepared (GREENBERG and EDELMAN 1983a). Proteins have been identified in several mammalian species which appear homologous to chicken p36 by peptide mapping and antigenicity, and which are phosphorylated at tyrosine under comparable conditions to chicken p36 (ERIKSON and ERIKSON 1980; COOPER and HUNTER 1981b; HUNTER and COOPER 1981; ERIKSON et al. 1981b; DECKER 1982).

p36 is phosphorylated at tyrosine in cells transformed by most of the viruses listed in Table 1 (Fig. 1; ERIKSON et al. 1981a; COOPER and HUNTER 1981b; HUNTER and COOPER 1983b; and our unpublished observations). It is not extensively phosphorylated in normal cells – whether resting, growing, infected with leukosis viruses (Fig. 1), or treated with tumor promoters (LASZLO et al. 1981) – or in cells transformed by chemicals, by DNA viruses, or by any of a host of retroviruses which appear to have no tyrosine protein kinase involvement – including MC29 virus, Moloney murine sarcoma virus, Harvey and Kirsten sarcoma viruses, and McDonough feline sarcoma virus (RADKE and MARTIN 1979; COOPER and HUNTER 1981b; CHENG and CHEN 1981; and our unpublished results). A slight increase in p36 phosphorylation following transformation by avian erythroblastosis virus has been reported (RADKE and MARTIN 1979). p36 is phosphorylated in some normal cells, after exposure to growth factors. pp36 is a major phosphoprotein of EGF-treated A431 cells, which have high levels of EGF receptors, and is detectable in some lines of human fibroblasts and mouse 3T3 cells after exposure to PDGF (HUNTER and COOPER 1981; COOPER and HUNTER 1981c; ERIKSON et al. 1981b; COOPER et al. 1982; BISHOP JM, personal communication). However, it is not phosphorylated in some other 3T3 cell lines after EGF or PDGF treatment, or in chicken embryo cells after mitogen stimulation, even though these cells are mitogenically responsive (COOPER et al. 1982; DECKER 1982; NAKAMURA et al. 1983). Thus it appears that p36 phosphorylation does not correlate with conditions of mitogenic stimulation. p36 phosphorylation has also been examined in chicken embryo cells infected with mutant strains of RSV (NAKAMURA and WEBER 1982; COOPER et al. 1983a). There is poor correlation between p36 phosphorylation and abrogation of normal restraints on cell growth, as assayed by the density at which monolayer cultures reach saturation or by the ability of cells to grow independently of a solid support. There is also poor correlation with decreased adhesion to the substratum, and a qualitative dissociation of cell morphology changes from p36 phosphorylation. There is, however, a significant correlation with one aspect of cell transformation, i.e., the secretion of extracellular protease able to activate plasminogen (NAKAMURA and WEBER 1982; COOPER et al. 1983a). The implications of this are unclear, and, as with all correlations, a mutant may be found which provides an exception. It is quite possible that p36 phosphorylation could correlate best with some aspect of transformation which is not readily quantified.

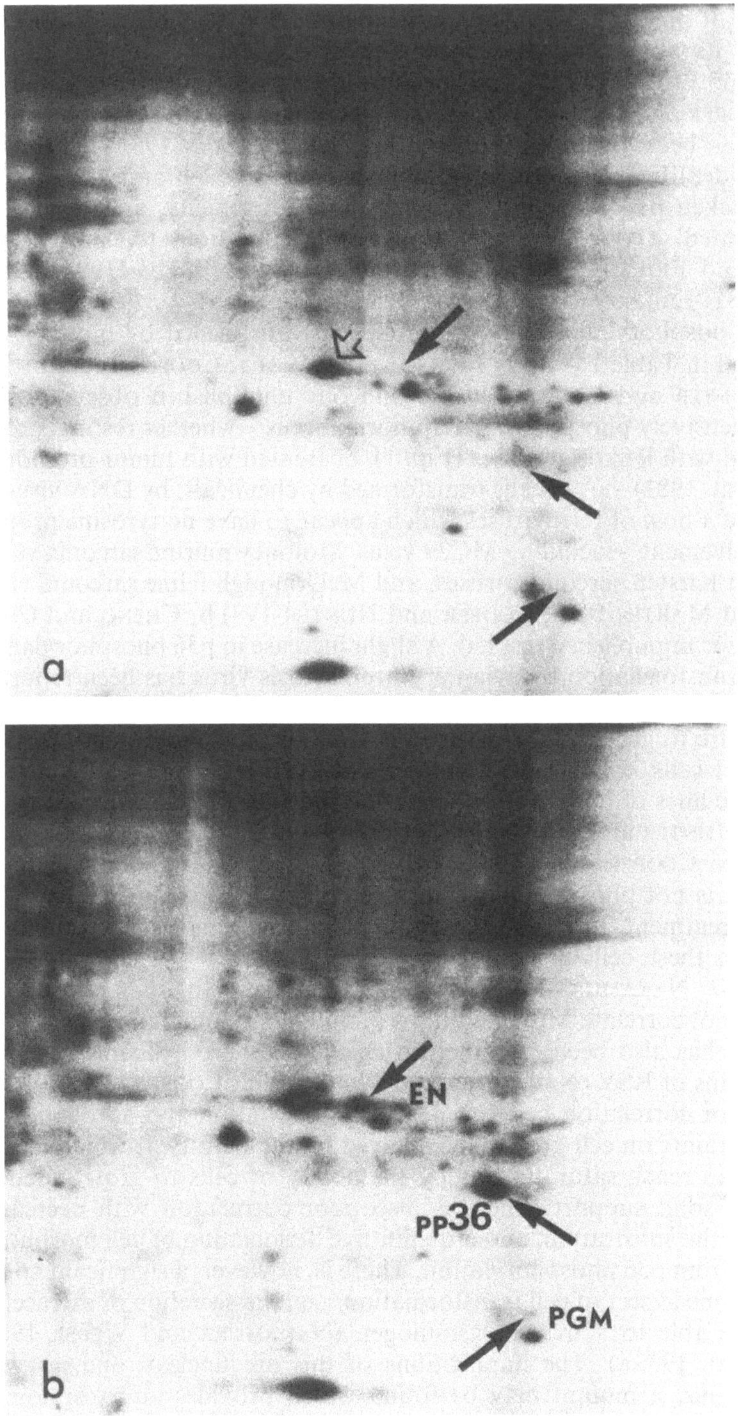
a
b
EN
PP36
PGM

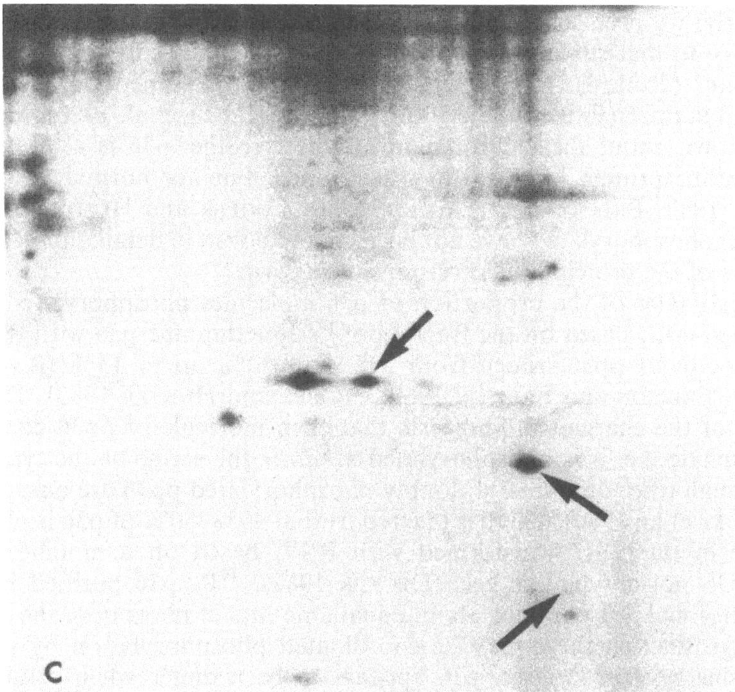

Fig. 1 a–c. Phosphoproteins of normal and transformed chicken embryo cells. Chicken embryo cells were infected with the amphotropic murine leukemia virus, 1504A, which can replicate in avian cells, and replicate cultures were **a** kept as control or **b** superinfected with Esh sarcoma virus or **c** superinfected with Y73 sarcoma virus. Phosphoproteins were labeled by incubation of the cells overnight with ^{32}P-orthophosphate, and separated by two-dimensional gel electrophoresis. The gels were incubated in alkali to facilitate detection of the phosphotyrosine-containing phosphoproteins previously identified (COOPER and HUNTER 1981 a, b). Autoradiographs are shown with the acidic end of the isoelectric focusing gel at the *left*. *Arrows* point to the positions of pp36 and phosphotyrosine-containing forms of enolase (*EN*) and phosphoglycerate mutase (*PGM*). The *open arrow* in **a** points to a constitutive, phosphorylated form of enolase which contains principally phosphoserine (see Sect. 2.5). (Experiment performed in collaboration with Tilo Patschinsky)

It appears that the same tyrosine residue is the principal phosphate acceptor both in cells transformed by various viruses and in EGF-treated A431 cells, and in vitro, when purified p36 is incubated with RSV pp60$^{v\text{-}src}$ (COOPER and HUNTER 1981 b, c; ERIKSON et al. 1981 a, b; ERIKSON and ERIKSON 1980). The ^{32}P-phosphorylated peptide from mouse pp36 has been sequenced and the ^{32}P-phosphotyrosine is released after 14 cycles of N-terminal degradation (PATSCHINSKY et al. 1982). Similar sequencing of a peptide generated with *S. aureus* V8 protease releases ^{32}P in cycle 10 (PATSCHINSKY et al. 1982), suggesting a partial sequence of the phosphorylation site:

lys) . x . x . x . *glu* . x . x . x . x . x . x . x . x . x . **tyr** . x
arg)

Clearly, we need more information to evaluate the relationship of this sequence to that surrounding the phosphorylation sites in viral transforming proteins (Table 3). A more minor phosphotyrosine peptide has also been found in pp36 (ERIKSON and ERIKSON 1980; RADKE et al. 1980), but nothing is known about the environment of this tyrosine. p36 is also a substrate for serine protein kinase(s) in transformed, but not normal, cells (RADKE et al. 1980; ERIKSON and ERIKSON 1980; COOPER and HUNTER 1981). The site(s) phosphorylated have not been characterized in detail, nor is the provenance of the protein kinase responsible known.

Estimates of the proportion of p36 molecules phosphorylated in transformed cells, based on the fraction of [^{35}S]methionine-p36 with the isoelectric point of pp36, range from less than 10% up to 15% (RADKE et al. 1980; ERIKSON and ERIKSON 1980; COOPER and HUNTER 1983). The magnitude of the charge shift suggests that each molecule of pp36 contains one phosphate, i.e. it is phosphorylated at *either* the serine *or* the tyrosine site, although trace amounts of doubly phosphorylated pp36 are also detectable (RADKE et al. 1980). There is one report that 40%–50% of p36 is phosphorylated in rat cells, transformed with RSV, based on a mobility decrease in SDS polyacrylamide gels (DECKER 1982). ^{32}P-pp36 purified from two-dimensional gels contains about equal amounts of phosphoserine and phosphotyrosine, so there may be coordinated phosphorylation by serine and tyrosine protein kinases. p36 appears to be a dimer when solubilized by non-ionic detergent and chelating agents, so it is possible that phosphorylation of one subunit at tyrosine could activate a susceptible serine(s) in the other subunit (ERIKSON and ERIKSON 1980). pp36, in chicken embryo cells transformed by a temperature-sensitive RSV, is dephosphorylated rapidly after inactivation of the pp60$^{v\text{-}src}$ by elevating the temperature (RADKE and MARTIN 1979; our unpublished data), suggesting that its phosphorylation may normally be limited by phosphatase action. Alternatively, the low stoichiometry of phosphorylation may reflect constraints on the kinase/substrate interaction. In normal cells, the proportion of phosphorylated p36 is very low, and phosphotyrosine has not been detected, suggesting that tyrosine protein kinases of normal cells may not recognize p36.

It is surprising that the function of such a major cell protein is unknown. It has been reported that a major 36-K phosphoprotein of RSV-transformed chicken embryo cells co-purifies with a cytosolic malate dehydrogenase activity (RUBSAMEN et al. 1982), but it is not clear whether the phosphoprotein is the same as the pp36 described by most investigators. Highly purified p36 preparations, albeit prepared from normal cells, lack malate dehydrogenase activity (COURTNEIDGE et al. 1983; GREENBERG and EDELMAN 1983b; ERIKSON RL, personal communication). Until the protein structure of this malate dehydrogenase is compared with p36, it is questionable whether p36 really has malate dehydrogenase activity.

Clues to the function of p36 have been sought from its subcellular location and from its level of expression in various differentiated cells. After cell fractionation by high-speed centrifugation in the presence of divalent cations, most p36 is in the particulate material, but in the absence of divalent cations about 30% is soluble (COOPER and HUNTER 1982). COURTNEIDGE et al. (1983) and GREENBERG and EDELMAN (1983b) have carefully analyzed the distribution of p36 in lysates containing divalent cations. Most p36 pellets at high speed together with membrane vesicles, not at low speed with nuclei. Analysis on discontinuous sucrose gradients shows that the p36-containing membranes are most likely plasma membrane (COURTNEIDGE et al. 1983; GREENBERG and EDELMAN 1983b; RADKE et al. 1983). pp36 is also associated with the particulate fraction of cells (COOPER and HUNTER 1982); specifically, with the plasma membrane (AMINI and KAJI 1983; COURTNEIDGE et al. 1983). Only a fraction of p36 can be released from the membranes by EDTA (COURTNEIDGE et al. 1983); p36 might thus be described as an "integral" membrane protein. However, it is not synthesized with a cleavable "signal" sequence, does not appear to be glycosylated or fatty-acylated, is a product of free, not membrane-bound ribosomes, and is not secreted (COURTNEIDGE et al. 1983; SEFTON B, personal communication; our unpublished observations). High salt concentrations effectively solubilize most p36, also suggesting that it is not an integral membrane protein (GREENBERG and EDELMAN 1983b). It is not degraded when cells are incubated with trypsin, implying that it is not exposed on the cell surface (COOPER and HUNTER 1982). Immunofluorescence of fixed, intact cells also shows that it is not accessible on the outer surface of the cell. However, when fixed cells are permeabilized to allow anti-p36 antibody to enter, fluorescence is associated with membrane ruffles and cell-cell junctions, which are regions of high plasma membrane concentration (NIGG et al. 1983; COURTNEIDGE et al. 1983; GREENBERG and EDELMAN 1983b; RADKE et al. 1983). p36 could be one of the most abundant proteins on the inner face of the fibroblast plasma membrane.

Localization of p36 to the plasma membrane is interesting in view of the concentration of most known tyrosine protein kinases in this vicinity (e.g., the growth factor receptors and pp60$^{v\text{-}src}$; WILLINGHAM et al. 1979; COURTNEIDGE et al. 1980). Endogenous p36 can be phosphorylated in crude in vitro systems consisting of A431 cell membranes and cytoplasm (GOSH-DASTIDAR and FOX 1983) in which the presumed tyrosine protein kinase is the EGF receptor, or in RSV-transformed cell lysates (KOBAYASHI et al. 1982) or membranes thereof (AMINI and KAJI 1983), in which the presumed tyrosine protein kinase is pp60$^{v\text{-}src}$. However, in vivo experiments suggest that subcellular localization may not in fact be of prime importance in substrate selection by tyrosine protein kinases. p36 is phosphorylated effectively in cells containing variant pp60$^{v\text{-}src}$ molecules which are not associated with the membrane (KRUEGER et al. 1982). Note also that there are examples of soluble cytoplasmic substrate proteins which are phosphorylated on tyrosine in RSV-transformed cells (Sect. 2.5).

Membrane proteins might be expected to be solubilized by detergents. However, when cultured fibroblasts are gently extracted with non-ionic detergent, most pp36 remains in the substrate-attached detergent-insoluble matrix (CHENG and CHEN 1981; COOPER and HUNTER 1982). This is consistent with the observation that after homogenization in the presence of nonionic detergent and centrifugation at high speed, most p36 is found in the pellet (GREENBERG and EDELMAN 1983b). However, there are two reasons for supposing that p36 is not a true component of the filamentous cytoskeleton. Firstly, if cells are detached from the substratum, gently extracted with detergent, and then centrifuged at low speed, p36 is in the supernatant, even though the subunit proteins of the intermediate filaments and microfilaments are pelleted under these conditions (COOPER and HUNTER 1982). Secondly, immunofluorescence of substrate-attached detergent-insoluble material shows that p36, and some other plasma membrane proteins, are in spongiform structures which probably represent vestiges of plasma membrane which have coalesced onto the submembrane cytoskeleton (Fig. 2; NIGG et al. 1983). In contrast, vinculin and several other cytoskeletal proteins are not disturbed by detergent extraction (Fig. 2; NIGG et al. 1983). A role for p36 in plasma membrane structure or function is indicated, possibly acting as a bridge between the membrane and cytoskeleton. In this light, it is interesting that p36 is extractable from detached cells, since in rounded cells, membrane-cytoskeletal interactions are weakened (PENMAN et al. 1981).

One very recent report (ARRIGO et al. 1983) suggests that a significant population of p36 is associated with ribonucleoprotein particles. This appeared to be a specific association, since p36 could be cross-linked to RNA before cell lysis. No striking differences between normal and transformed cell p36 were detected, however.

Even though p36 is abundant in fibroblasts, it is not found in all cells. Some lines derived from pre-B lymphomas contain p36 at less than 1% of its abundance in fibroblasts; some other lymphoid lines have a p36 content similar to fibroblasts (SEFTON et al. 1983). The limit of detection in these experiments was about 2×10^4 molecules per cell (assuming a protein content of 7×10^{-11} g/lymphoid cell). Clearly, high abundance of p36 is not essential for cell viability. Remarkably, some of these p36-deficient lines are transformed by AMuLV, suggesting that the presence of pp36 is not essential for lymphoid cell transformation, although it correlates well with the transformation of fibroblasts (SEFTON et al. 1983). The abundance of p36 is also low in some animal tissues (GOULD K, unpublished results; ERIKSON RL, personal communication). It is barely detected by immunological techniques in brain, liver, and skeletal muscle; is present at intermediate concentrations in heart, kidney, spleen, and testis; and is at its highest

Fig. 2. Subcellular localization of p36 in detergent-treated NRK cells. NRK cells were briefly ▷ treated with a solution containing non-ionic detergent before formaldehyde fixation. p36 (*upper*) and vinculin (*lower*) were visualized by reaction with rabbit and guinea pig antibodies, respectively, followed by rhodamine- and fluorescein-conjugated second antibodies. All procedures are described in NIGG et al. (1983). Note the reticular p36 fluorescence compared with the punctate fluorescence of vinculin, which is in adhesion plaques. The nuclear fluorescence with the anti-p36 serum could be prevented by preabsorption of this serum on lymphoid cells which lack p36. The *bar* in the *upper panel* denotes 20 μm. (Experiment performed by Erich Nigg)

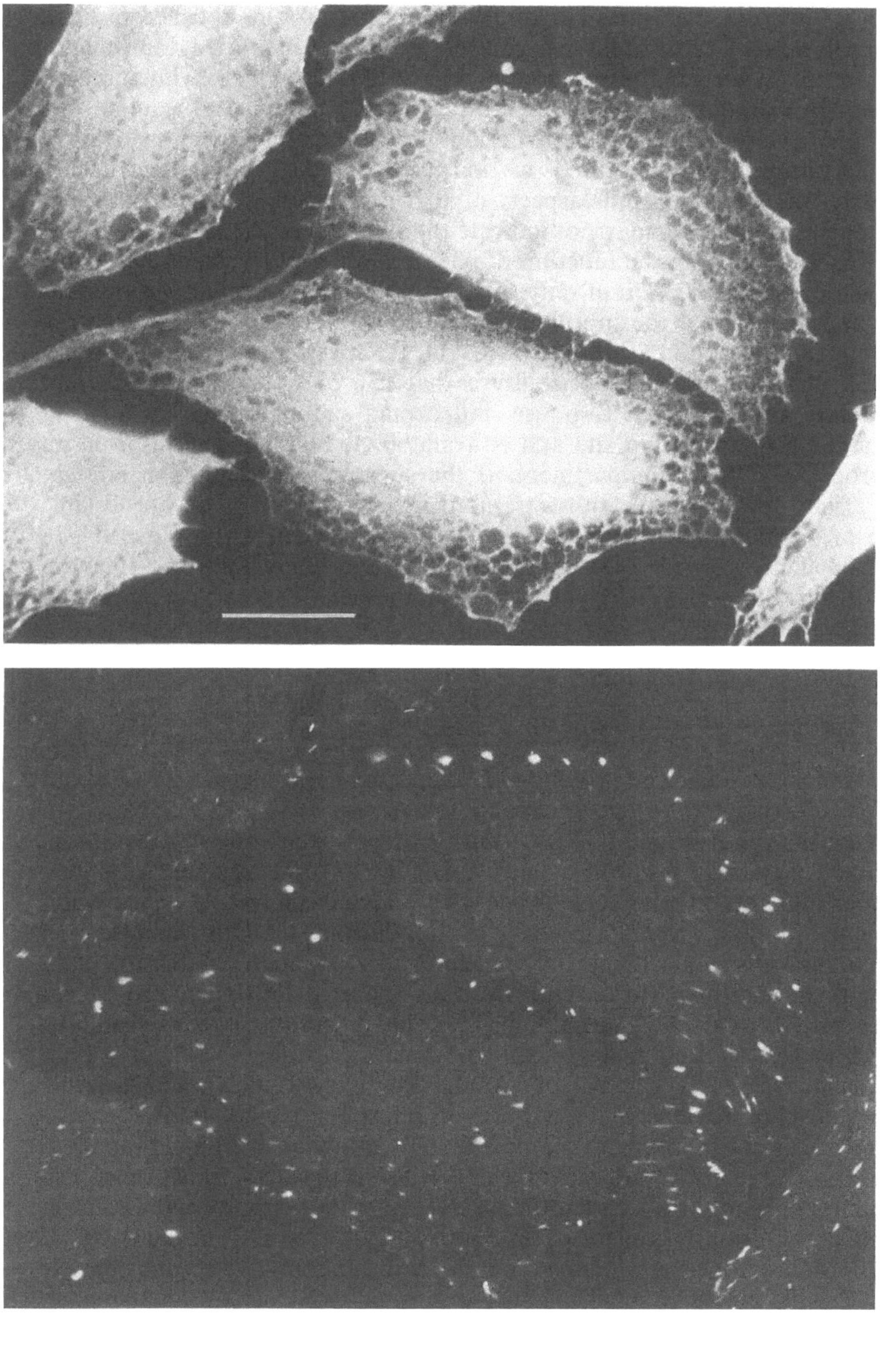

level in thymus, small intestine, and lung (GOULD K, unpublished results). The p36-rich cell types in these organs have not been identified, but preliminary results suggest that the structural cells of the spleen and thymus contain more p36 than do the lymphoid cells (GOULD K, unpublished results). The low p36 level in adult muscle is curious in view of the detection of pp36 in RSV-transformed cultured myotubes (KOBAYASHI and KAJI 1980). Clearly, the developmental aspects of p36 expression will be interesting, although these studies may not indicate the function of the protein.

Without knowing the function of p36, it is difficult to assess the significance of phosphorylation of only 15% of molecules. Phosphorylation does not appear to affect the subcellular location of p36 (COOPER and HUNTER 1982; COURTNEIDGE et al. 1983; NIGG et al. 1983; GREENBERG and EDELMAN 1983b). One unattractive possibility is that p36 phosphorylation does not modulate any biological response, but occurs simply because p36 has a suitable phosphorylation site and is available at high concentration in an appropriate cellular compartment. If the substrate concentration is high, the affinity of kinase for substrate need not be strong for an appreciable rate of reaction. The affinity of the reaction in vitro, using purified p36 and pp60$^{v\text{-}src}$, has not been reported. When we understand the function of p36, we will be in a better position to test the significance of the phosphorylation.

2.4 Vinculin

Vinculin is a cytoskeletal protein with a reported molecular weight of about 130 K (GEIGER 1979; BURRIDGE and FERAMISCO 1980). It was found to contain phosphotyrosine in RSV-transformed chicken embryo cells during a survey of cytoskeletal proteins undertaken by SEFTON et al. (1981a). Transformed and normal cells were labeled with $^{32}P_i$ and cell lysates were precipitated with antisera to myosin heavy chain, vinculin, filamin, vimentin, and α-actinin. Nonspecifically precipitated actin was also analyzed (SEFTON et al. 1981a). Extracellular fibronectin has been analyzed by similar procedures (ALI and HUNTER 1981). Vinculin was the only one of the proteins examined which proved to contain significant amounts of phosphotyrosine. The bulk of the phosphate in vinculin of normal cells is present as phosphoserine and phosphothreonine, with a trace of phosphotyrosine. In RSV-transformed cells phosphotyrosine comprises 20% of phosphate linked to vinculin. Approximately 1% of vinculin molecules in the RSV-transformed cell contain phosphotyrosine, a tenfold increase over the proportion phosphorylated at tyrosine in control cells (SEFTON et al. 1981a). The majority of the phosphotyrosine is contained in a single phosphorylated tryptic peptide, which is detectable in small quantity in normal cell vinculin (SEFTON et al. 1981a).

Vinculin is phosphorylated to a similar extent in chicken embryo cells transformed by Y73, but to a much lesser extent in cells transformed by PRCII or FSV (SEFTON et al. 1981a, c). It is therefore more likely that vinculin is phosphorylated directly by the respective viral transforming proteins than by a cellular tyrosine protein kinase which is invariably activated as a consequence of transformation. Only a few cell lines transformed by mammalian retroviruses have been tested for phosphorylated vinculin. However, it appears that vinculin is phosphorylated at tyrosine in mouse fibroblasts transformed by AMuLV and RSV but not by SV40 (SEFTON et al. 1981a, c). Vinculin appears not to be extensively phosphorylated at tyrosine in quiescent mouse 3T3 cells in response to PDGF (COOPER et al. 1982), and vinculin phosphorylation does not vary during the cell cycle (ROSOK and ROHRSCHNEIDER 1983). The low constitutive level of phosphotyrosine in normal cell vinculin implies that this protein is a substrate for tyrosine protein kinase(s) found in normal cells (SEFTON et al. 1981a).

Teleologically, it was likely that at least one cytoskeletal protein was a substrate for viral tyrosine protein kinases, since the cytoskeleton is dramatically restructured by transformation. The discovery that vinculin phosphorylation accompanied transformation by some retroviruses prompted speculation that this phosphorylation may trigger the cytoskeletal changes. About 20% of cellular vinculin is found in normal cells concentrated in adhesion plaques, where the outer surface of the membrane is bonded to the extracellular matrix and the inner surface provides an anchorage point for stress fibres composed of actin microfilaments (see Fig. 2; GEIGER 1979; BURRIDGE and FERAMISCO 1980). Both the total cellular microfilament system (ASH et al. 1976; EDELMAN and YAHARA 1976; WANG and GOLDBERG 1976) and the adhesion plaques are reorganized in transformed cells. Both vinculin and pp60$^{v\text{-}src}$ are now found in vestiges of adhesion plaques, where the plasma membrane approximates to the substratum (DAVID-PFEUTY and SINGER 1980; ROHRSCHNEIDER 1980; SHRIVER and ROHRSCHNEIDER 1981; SEFTON et al. 1982; NIGG et al. 1982). In RSV-transformed cells exhibiting tight junctions, both vinculin and pp60$^{v\text{-}src}$ are found together in these structures (WILLINGHAM et al. 1979; SHRIVER and ROHRSCHNEIDER 1981). Thus a reasonable hypothesis is that pp60$^{v\text{-}src}$, present in adhesion plaques and tight junctions, could phosphorylate a fraction of vinculin molecules at tyrosine and interfere with interactions of these molecules with other adhesion plaque components, including the microfilament system, possibly in a cooperative fashion (for a graphic illustration see HUNTER 1982b). Whether the tyrosine-phosphorylated vinculin remains associated with the adhesion plaques or is released is unknown, but phosphorylated vinculins present in normal cells are predominantly located in the cytoskeletal fraction (GEIGER 1982). The failure to detect vinculin phosphorylation, at tyrosine, in PRCII-transformed cells, could be fitted into the scheme since these cells are fusiform; a morphology which is indicative of incomplete morphological transformation (SEFTON et al. 1981a).

Clearly, however, the phosphorylation of vinculin is not the sole basis for the changes in cell shape which accompany transformation, and alteration of the amount or structure of other molecules is probably important. Cells transformed by agents which do not demonstrably affect protein phosphorylation at tyrosine are morphologically altered yet do not contain high levels of phosphotyrosine in their vinculin. Examination of vinculin phosphorylation in cells infected with mutants of RSV shows that vinculin tyrosine phosphorylation is not sufficient for a rounded morphology. The RSV mutant F1 induces a fusiform morphology, like that of cells transformed by PRCII virus, despite inducing noticeable phosphorylation of vinculin at tyrosine (SEFTON et al. 1981c). In a wide survey of RSV mutants, ROHRSCHNEIDER has found no correlation between the phosphorylation state of vinculin and (a) gross cellular morphology or (b) the localization of $pp60^{v-src}$ (ROHRSCHNEIDER et al. 1981; ROHRSCHNEIDER and ROSOK 1983). An exception to the apparent correlation between vinculin phosphorylation on tyrosine and loss of stress fibers (ROHRSCHNEIDER et al. 1981) was found when another RSV mutant, RD10, was studied (ROHRSCHNEIDER and ROSOK 1983). Cells infected with this mutant do contain stress fibers, but their vinculin is phosphorylated at tyrosine. These data are consistent with vinculin phosphorylation possibly being necessary, but clearly being insufficient, to induce the dissolution of stress fibers (ROHRSCHNEIDER and ROSOK 1983). In addition, mitosis in normal cells is a situation where increased vinculin phosphorylation is not necessary for release of stress fibers from adhesion plaques (ROSOK and ROHRSCHNEIDER 1983). Apparently there is no simple relationship between vinculin phosphorylation and stress fiber organization. Possibly biochemical localization of those vinculin molecules containing phosphotyrosine, by cell fractionation, will be more informative than comparison of immunofluorescent localization studies with phosphoamino acid analysis of bulk vinculin.

Purified chicken gizzard vinculin is phosphorylated in vitro, by a purified fragment of $pp60^{v-src}$ (ITO et al. 1982). Curiously, this reaction is stimulated by phospholipid; specifically, by anionic phospholipids (ITO et al. 1982). Even though vinculin is not thought to be an integral membrane protein, the phospholipid probably interacts with vinculin rather than with the protein kinase, since the phosphorylation of other added substrates is not stimulated (ITO et al. 1982). Moreover, the protein kinase used in these studies was a naturally occurring proteolytic fragment of $pp60^{v-src}$ which has lost its presumed membrane location site (BLITHE et al. 1982; LEVINSON et al. 1981). Provided that the sites phosphorylated in vitro are the same as those phosphorylated in vivo, then this system may provide a route to studying the effect of phosphorylation on vinculin function, since vinculin may be assayed in vitro for its ability to inhibit the gelation of actin solutions.

2.5 Glycolytic Enzymes

Phosphorylation of 46-K and 28-K proteins at tyrosine was first observed in RSV-transformed chicken embryo cells during analysis of ^{32}P-labeled cells by two-dimensional gel electrophoresis (COOPER and HUNTER 1981a). They were subsequently shown to be phosphorylated, also at tyrosine, in chicken cells transformed by FSV, PRCII, Y73, and ESV (see Fig. 1; COOPER and HUNTER 1981b). 46-K and 29-K phosphoproteins were found in

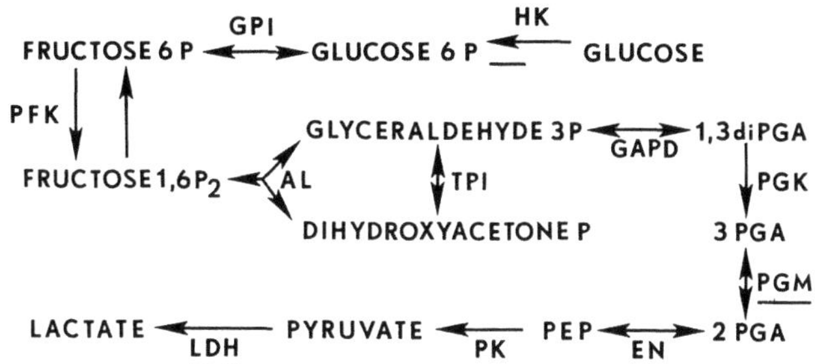

Fig. 3. Summary of the reactions of glycolysis. The glycolytic enzymes, in order, are: *HK*, hexokinase; *GPI*, glucose phosphate isomerase; *PFK*, phosphofructokinase; *AL*, aldolase; *TPI*, triose phosphate isomerase; *GAPD*, glyceraldehyde phosphate dehydrogenase; *PGK*, phosphoglycerate kinase; *PGM*, phosphoglycerate mutase; *EN*, enolase; *PK*, pyruvate kinase; *LDH*, lactate dehydrogenase. No cofactors are shown. Enzymes which are phosphorylated at tyrosine in RSV-transformed cells are underlined

mouse 3T3 cells transformed by RSV, AMuLV (46-K protein not phosphorylated extensively), and STFeSV, and have since been shown to be homologous to the chicken proteins (Cooper and Hunter 1981 b; Hunter and Cooper 1983 b; our unpublished observations). The STFeSV-transformed mouse cells also contained significant amounts of a 35-K phosphoprotein, which, like the others, contained phosphotyrosine and phosphoserine. None of these phosphoproteins were detected in normal chicken or 3T3 cells, even when treated with growth factors, or in 3T3 cells transformed with SV40, polyoma virus, Moloney murine sarcoma virus, or McDonough feline sarcoma virus (Cooper and Hunter 1981 a, b; Cooper et al. 1982; our unpublished observations). The significance of these anonymous proteins is now evident since the 46-K, 28- to 29-K, and 35-K proteins have been identified, by immunoprecipitation and peptide mapping, as the glycolytic enzymes enolase, phosphoglycerate mutase, and lactate dehydrogenase, respectively (Fig. 3; Cooper et al. 1983 b).

Examination of the enzymes aldolase, triose phosphate isomerase, glyceraldehyde phosphate dehydrogenase, phosphoglycerate kinase, and pyruvate kinase, immunoprecipitated from RSV-transformed cells, revealed that they were not significantly phosphorylated at tyrosine, despite, in some cases, being present in the cell in equal or greater abundance than the three enzymes which were phosphorylated (Cooper et al. 1983 b). Incubation of ^{32}P-labeled extracts of RSV-transformed or normal chicken cells with antisera to hexokinases precipitated proteins which were not phosphorylated (Sefton BM, personal communication).

The three glycolytic enzymes which are tyrosine protein kinase substrates resemble p36 in that they are abundant cell proteins and only a small proportion of the population (5%–10%) becomes phosphorylated in transformed cells (Cooper and Hunter 1983). Since the enzymes are oligomers (enolase and phosphoglycerate mutase are dimers, lactate dehydrogenase is a tetramer) the proportion of phosphorylated enzymes molecules could approach

20%. Even so, there is a large population of each enzyme which is not phosphorylated, so there is the possibility that the phosphorylation is inconsequential. Of course, if the fraction of molecules which is phosphorylated comprises a critical population actively involved in glycolysis and/or gluconeogenesis, then the phosphorylations could be disruptive to cell metabolism.

Unlike p36, our knowledge of the functions of these substrates means that it should be possible to examine whether phosphorylation alters their activity by using in vitro assays. This is of special interest in view of the notorious increased rate of aerobic glycolysis in many tumor cells, known as the Warburg effect (WARBURG 1930, see also RACKER 1972). However, cells transformed in many ways, and not only by viruses encoding tyrosine protein kinases, exhibit the Warburg effect, yet their enolase and phosphoglycerate mutase, and presumably lactate dehydrogenase, are not phosphorylated at tyrosine (COOPER and HUNTER 1981a, b). Phosphorylation of these glycolytic enzymes at tyrosine would appear not to be the only possible cause of, and may in fact be unrelated to, the Warburg effect. Since the reactions they catalyze lie at, or close to, thermodynamic equilibrium in the cell as a whole, activation of these enzymes will not alter the relative concentrations of their substrates and products (Fig. 3). Indeed, when levels of glycolytic intermediates were measured in RSV-transformed chicken embryo cells and compared with those in normal cells, the ratio of substrate to product concentration for enolase, phosphoglycerate mutase, and lactate dehydrogenase were little altered (BISSELL et al. 1973; SINGH et al. 1974). Instead, the enzymes whose substrates were present at reduced relative levels in transformed cells were phosphofructokinase and pyruvate kinase, implying that these two enzymes were activated by transformation (BISSELL et al. 1973; SINGH et al. 1974). Pyruvate kinase appears not to be phosphorylated at high level at tyrosine in RSV-transformed cells (COOPER et al. 1983b), even though it can be phosphorylated by $pp60^{v\text{-}src}$ in vitro (PRESEK et al. 1980). The phosphoamino acid composition of transformed-cell phosphofructokinase has not been examined. Another important control point is hexose transport. The increased maximal velocity of hexose transport in RSV-transformed cells (WEBER 1973) could supply the drive for increased glycolysis.

These findings do not encourage a belief that tyrosine phosphorylation contributes to altered carbohydrate metabolism. Nevertheless, the interactions between individual glycolytic enzymes, and individual gluconeogenic enzymes, are exceedingly complex, since the key regulatory enzymes are controlled by the concentrations of many effectors. Moreover, it is possible that glycolytic enzymes are spatially organized in the cell (SIGEL and PETTE 1969; CLARKE and MASTERS 1974; FULTON 1982), and phosphorylation could alter this organization. Hence cooperative interactions could amplify the effects of a small degree of phosphorylation. Within such a complex, there is no reason why the reversible steps of glycolysis should lie at equilibrium, and the reaction catalyzed by, say, enolase, could be rate limiting.

It is also possible to imagine that phosphorylation of some enzymes could alter the rates of reactions catalyzed by other enzymes in the complex.

Very little is known concerning the phosphorylation sites in these substrates. Before its identity was known, the 46-K protein was purified from chicken embryo cells, and an antiserum obtained (COOPER and HUNTER 1983). Immunoprecipitated 46-K protein of normal cells contains largely phosphoserine, whereas that from RSV-transformed cells contains equal amounts of phosphotyrosine and phosphoserine (COOPER and HUNTER 1983). There is a single major serine phosphorylation site in enolase from uninfected cells. Transformed cell enolase contains a second major phosphopeptide, which contains phosphotyrosine, and a third, minor peptide also containing phosphotyrosine (our unpublished data). The phosphorylated enolase found in normal as well as transformed cells can be distinguished from the species unique to transformed cells by its more acidic isoelectric point (see Fig. 1; COOPER and HUNTER 1981 a). This more acidic, constitutive form contains chiefly phosphoserine and some phosphothreonine. The more basic, conditional form contains phosphotyrosine and phosphoserine (COOPER and HUNTER 1981 a). The reason for the different isoelectric points of the constitutive and conditional enolase phosphoproteins is unclear, but each may be the phosphorylated form of a different nonphosphorylated enolase variant. There are indeed two nonphosphorylated enolase species in fibroblasts, of which the more basic species is more abundant (COOPER and HUNTER 1983). Presumably this latter variant is the substrate for tyrosine as well as serine protein kinases, in transformed cells only, and contains the two new phosphopeptides detected in transformed cell enolase. Phosphoserine can also be detected in a significant fraction of normal cell phosphoglycerate mutase and lactate dehydrogenase molecules, but in these cases the conditional species cannot be separated from the constitutive species by isoelectric focusing (COOPER and HUNTER 1981 a, 1983). Presumably phosphoglycerate mutase also contains phosphohistidine, since this is implicated in the catalytic mechanism, but this phosphoamino acid is acid labile and so is not detected in our routine analysis (COOPER et al. 1983 b).

As a preliminary to assaying the effects of phosphorylation on kinetic parameters, we have incubated purified enolase, phosphoglycerate mutase, and lactate dehydrogenase together with ATP and immunoprecipitated transforming proteins known to possess associated tyrosine protein kinase activity. In each case the glycolytic enzyme becomes phosphorylated at tyrosine (our unpublished data). The peptide phosphorylated in enolase is identical to the major phosphotyrosine containing peptide unique to transformed cell enolase. The peptides phosphorylated in phosphoglycerate mutase in vitro include those phosphorylated in transformed cells. The same single tyrosine residue is phosphorylated in lactate dehydrogenase in vitro and in vivo. Further experiments are necessary to determine the affinity of the kinase(s) for substrates, and to characterize the phosphorylation sites in detail.

2.6 Other Substrates

Several other substrates for tyrosine protein kinases are known only by their electrophoretic characteristics. They include 42-K to 45-K proteins whose phosphorylation is induced in quiescent 3T3 (COOPER et al. 1982), chick (NAKAMURA et al. 1983), or human (Fig. 4) cells by brief exposure to growth factors, and an 81-K protein phosphorylated in 3T3, mink lung, and A431 cells following transformation by STFeSV (HUNTER and COOPER 1983b; our unpublished results). The 81-K protein is also phosphorylated at tyrosine and serine in uninfected A431 cells following exposure to EGF (Fig. 5; HUNTER and COOPER 1981; COOPER and HUNTER 1981c). The conditions under which these proteins are phosphorylated leads us to hope that their phosphorylation will prove to be important, but since the substrates have yet to be purified, or their identities otherwise established, much further work is necessary. Additional tyrosine protein kinase substrates are currently being detected by novel techniques. In particular, some phosphoproteins are specifically retained on affinity columns prepared with antisera raised to phosphotyrosine analogues (ROSS et al. 1981). There are indications that monoclonal antibodies to phosphotyrosine may increase the sensitivity of this technique and enable the retention of phosphoproteins not detected so far by other approaches (FRACKELTON et al. 1983). It is tempting to speculate that some proteins phosphorylated at tyrosine may be found in the nucleus, there regulating gene expression.

3 One Critical Substrate or Many?

It is clear that we are still a long way from demonstrating the involvement of any single tyrosine phosphorylation event in the regulation of cell growth or transformation. In part this failure to demonstrate causality is due to the absence of suitable mutants. If phosphorylation of a particular substrate were important for, say, the ability to grow in soft agar, then a mutant cell type in which this protein was altered might not be transformable by viruses encoding tyrosine protein kinases. Such cells could be selected as revertants in the transformed cell population. However, revertant transformed cell mutants obtained to date do not have altered cellular genes but instead have all proved to be defective in the viral transforming protein; either it is mutated or the viral genome is deleted.

Possibly the absence of selectable "nontransformed" revertants, mutated in cellular genes, means that such mutations are lethal, or it could reflect a requirement for simultaneous mutations in more than one cellular gene to produce the revertant phenotype. It has been argued that the partial dissociation of some parameters of transformation by mutations in a viral genome means that the transforming protein interacts with several important targets, each of which is relevant to a different aspect of transformation

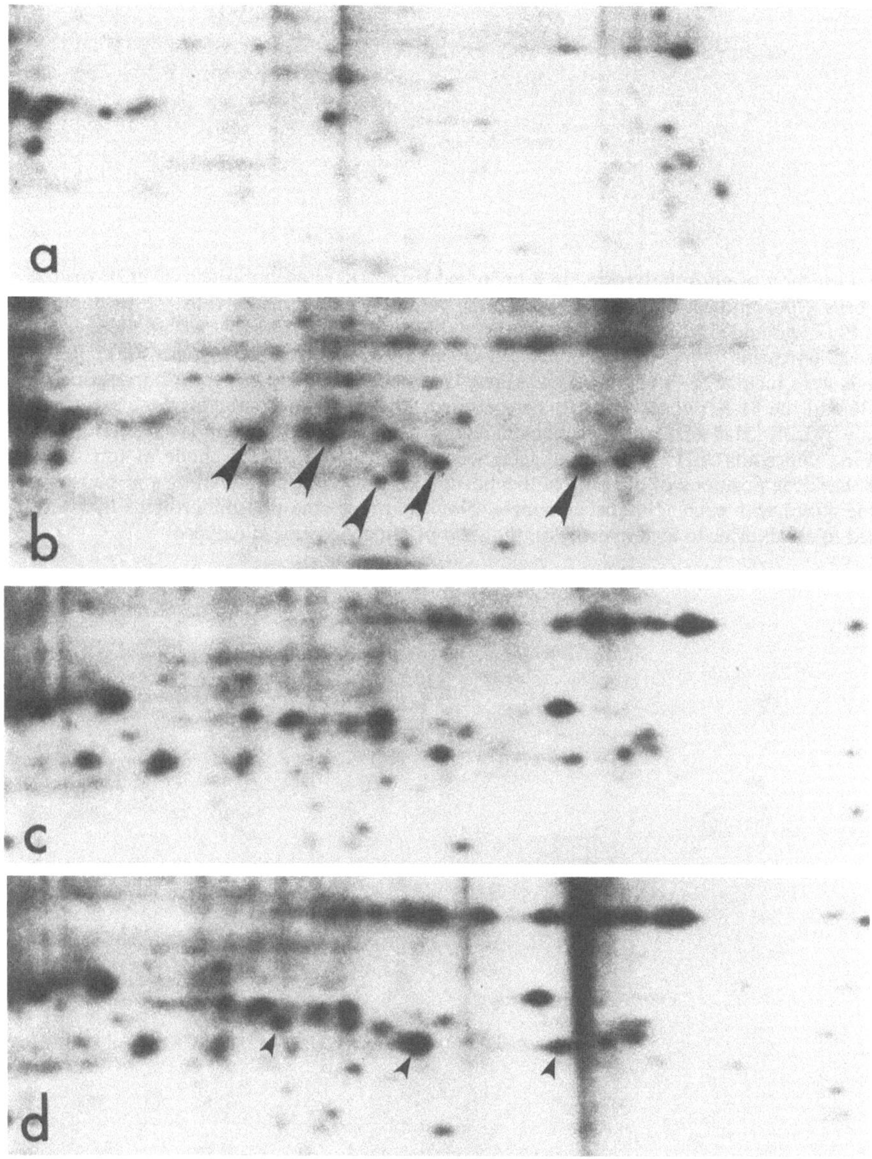

Fig. 4a–d. 42-K to 45-K phosphoproteins of PDGF-treated fibroblasts. Confluent cultures of NR-6 3T3 mouse cells (**a, b**) and AG1523 human diploid fibroblasts (**c, d**) were incubated with ^{32}P-orthophosphate overnight. One culture of each was kept as control (**a, c**) and the other treated with 0.33 n*M* pure PDGF for the final 1 h of labeling (**b, d**). Phosphoproteins were separated by two-dimensional gel electrophoresis, and the gels were incubated in alkali before autoradiography. For details, see Cooper et al. (1982). The region of each gel encompassing protein molecular weights in the range 35–65 K is shown. *Large arrowheads* in **b** indicate phosphoproteins whose phosphorylation is increased by PDGF in NR-6 3T3 cells. They contain phosphotyrosine (Cooper et al. 1982). *Small arrowheads* in **d** indicate homologous AG1523 cell phosphoproteins. The phosphoamino acids of these proteins have not been analyzed

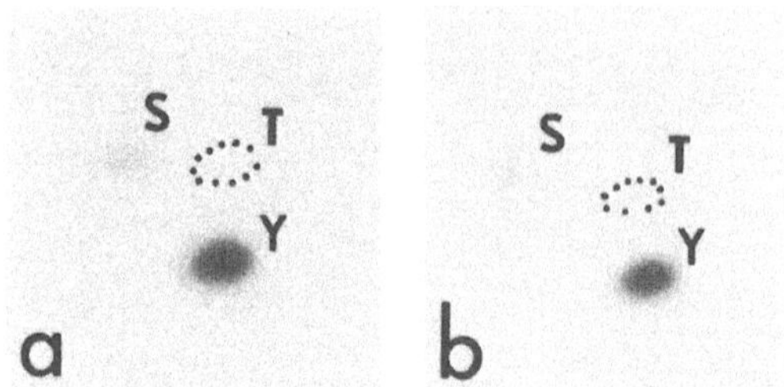

Fig. 5. Detection of phosphotyrosine in **a** pp36 and **b** an 81K phosphoprotein of EGF-treated A431 cells. Subconfluent cultures of A431 cells were labeled overnight with ^{32}P-orthophosphate. EGF was added to a concentration of 50 ng/ml 3 h before lysing the cells and separating the phosphoproteins by two-dimensional gel electrophoresis (HUNTER and COOPER 1981). Replicate gels were incubated in alkali (COOPER and HUNTER 1981a) and the spots corresponding to pp36 and the 81-K phosphoprotein were excised. Phosphorylated proteins were extracted, partially hydrolyzed in acid, and their phosphoamino acids separated by thin-layer electrophoresis. First dimension: pH 1.9 anode at left. Second dimension: pH 3.5 anode at top. *S*, *T*, and *Y* mark the positions of nonradioactive phosphoserine, phosphothreonine, and phosphotyrosine mixed with each ^{32}P-labeled sample. Note that since the phosphoproteins had been exposed to alkali prior to acid hydrolysis, the yield of phosphoserine is reduced

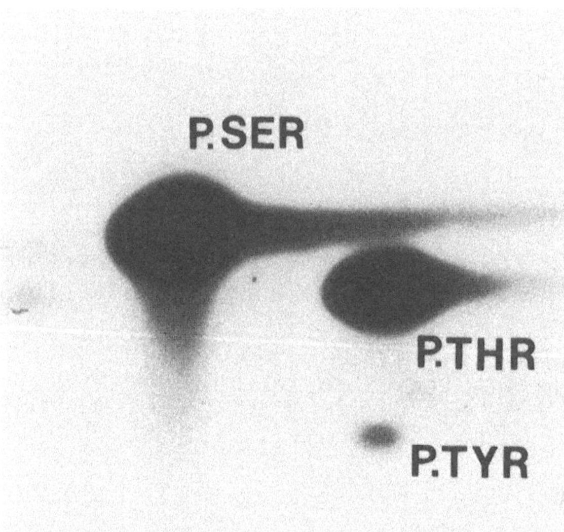

Fig. 6. Phosphotyrosine is present in *Drosophila* cell proteins. Mei 218A cells (courtesy of J. SIEGEL) were labeled overnight in 1 ml of complete Schneider's medium lacking ^{31}P-phosphate but supplemented with 7.6% fetal calf serum, 2.4% chicken serum, and 1 mCi ^{32}P-orthophosphate. The cells were lysed; total proteins were separated from nucleic acids, lipids, and low molecular weight material and then partially hydrolyzed in acid (procedure described in SEFTON et al. 1980b). Phosphoamino acids were separated by thin-layer electrophoresis, as described in the legend to Fig. 5. Of the radioactivity recovered in phosphoamino acids, 86% was in phosphoserine (*P.SER*), 14% in phosphothreonine (*P.THR*), and 0.07% in phosphotyrosine (*P.TYR*). (Experiment performed by Bryan Crenshaw)

(WEBER and FRIIS 1979). Probably the situation is more complicated. Each observable phenotypic change may be the result of *many* molecular changes.

Protein kinases provide a mechanism for the modification of many proteins in a cell; they permit both the amplification of a signal and its transmission to several receivers. Possibly many of the substrates for tyrosine protein kinases are important for each phenotypic change. In this case a genetic approach is doomed. It may be more profitable to carry out studies with simpler organisms for which genetic manipulations are routine. Since some of the tyrosine protein kinase genes have been recognized in *Drosophila* (SHILO and WEINBERG 1981), and there is phosphotyrosine in proteins of *Drosophila* (Fig. 6), protozoans, and bacteria (MANAI and COZZONE 1982; CRENSHAW B, HUNTER T, LEVY B, personal communications), tyrosine protein phosphorylation may play a role in inter- or interacellular regulation or communication in lower organisms. Identification of a metabolic or behavioral change correlating with tyrosine protein kinase expression may be predicted. Of course, it is quite possible that several substrates will mediate even the "simple" response of "simple" organisms, but otherwise genetic analysis may be used to dissect the chain of events. It is questionable, however, how such studies would relate to the problems of growth control and malignant transformation in vertebrate cells.

Acknowledgements. We thank our collaborators, BART SEFTON, ERICH NIGG, KATHY GOULD, TILO PATSCHINSKY, and BRYAN CRENSHAW, for permission to quote and show unpublished data, and our colleagues in other laboratories for their free communication of results before publication. BART SEFTON, CHRIS SARIS, and TOM CURRAN made helpful comments on the manuscript. This review would have taken even longer to write without the help of VAX. This work was supported by Public Health Service grants CA14195, CA17096, and CA28458.

References

Adkins B, Hunter T, Beemon K (1982a) Expression of the PRC II avian sarcoma virus genome. J Virol 41:767–780

Adkins B, Hunter T, Sefton BM (1982b) The transforming proteins of PRCII virus and Rous sarcoma virus form a complex with the same two cellular proteins. J Virol 43:448–455

Ali I, Hunter T (1981) Structural components of fibronectins from normal and transformed cells. J Biol Chem 256:7671–7677

Amini S, Kaji A (1983) Association of pp36, a phosphorylated form of the presumed target protein for the *src* protein of Rous sarcoma virus, with the membrane of chicken cells transformed by Rous sarcoma virus. Proc Natl Acad Sci USA 80:960–964

Arrigo A-P, Darlix J-L, Spahr P-F (1983) A cellular protein phosphorylated by the avian sarcoma virus transforming gene product is associated with ribonucleoprotein particles. EMBO Journal 2:309–315

Ash JF, Vogt PK, Singer SJ (1976) Reversion from transformed to normal phenotype by inhibition of protein synthesis in rat kidney cells infected with a temperature-sensitive mutant of Rous sarcoma virus. Proc Natl Acad Sci USA 73:2047–2051

Avruch J, Nemenoff RA, Blackshear PJ, Pierce MW, Osathanodh R (1982) Insulin-stimulated tyrosine phosphorylation of the insulin receptor in detergent extracts of human placental membranes. J Biol Chem 257:15162–15166

Barbacid M, Beemon K, Devare SG (1980) Origin and functional properties of the major gene product of the Snyder-Theilen strain of feline sarcoma virus. Proc Natl Acad Sci USA 77:5158–5162
Barbacid M, Breitman ML, Lauver AV, Long LK, Vogt PK (1981a) The transformation-specific proteins of avian (Fujinami and PRC-II) and feline (Snyder-Theilen and Gardner-Arnstein) sarcoma viruses are immunologically-related. Virology 110:411–419
Barbacid M, Donner L, Ruscetti SK, Scherr CJ (1981b) Transformation defective mutants of Snyder-Theilen feline sarcoma virus lack tyrosine specific protein kinase activity. J Virol 39:246–254
Beemon K (1981) Transforming proteins of some feline and avian sarcoma viruses are related structurally and functionally. Cell 24:145–153
Beemon K, Ryden T, McNelly EA (1982) Transformation by avian sarcoma viruses leads to phosphorylation of multiple cellular proteins on tyrosine residues. J Virol 42:742–747
Bishop JM, Varmus H (1982) Functions and origins of retroviral transforming genes. In: Weiss R, Teich N, Varmus H, Coffin J (eds) RNA tumor viruses. Cold Spring Harbor, New York, pp 999–1108
Bissell MJ, White RC, Hatie C, Bassham JA (1973) Dynamics of metabolism of normal and virus-transformed chick cells in culture. Proc Natl Acad Sci USA 70:2951–2955
Blithe DL, Richert N, Pastan IH (1982) Purification of a tyrosine-specific protein kinase from Rous sarcoma virus-induced rat tumors. J Biol Chem 257:7135–7142
Brautigan DL, Bornstein DL, Gallis B (1981) Phosphotyrosyl-protein phosphatase: specific inhibition by Zn^{2+}. J Biol Chem 256:6519–6522
Brugge J, Darrow D (1982) Rous sarcoma virus-induced phosphorylation of a 50,000 molecular weight cellular protein. Nature 295:250–253
Brugge J, Erikson E, Erikson RL (1981) The specific interaction of the Rous sarcoma virus transforming protein, $pp60^{src}$, and two cellular proteins. J Virol 26:773–782
Brugge J, Yonemoto W, Darrow D (1983) Interaction between the Rous sarcoma virus transforming protein and two cellular phosphoproteins: analysis of the turnover and distribution of this complex. Mol Cell Biol 3:9–19
Buhrow SA, Cohen S, Stavros JV (1982) Affinity labeling of the protein kinase associated with the epidermal growth factor receptor in membrane vesicles from A431 cells. J Biol Chem 257:4019–4022
Burridge KM, Feramisco J (1980) Microinjection and localization of a 130K protein in living fibroblasts: a relationship to actin and fibronectin. Cell 19:587–595
Carpenter G, King L Jr, Cohen S (1979) Rapid enhancement of protein phosphorylation in A-431 cell membrane preparations by epidermal growth factor. J Biol Chem 254:4884–4891
Casnellie JE, Harrison ML, Hellstrom KE, Krebs EG (1982a) A lymphoma protein with an in vitro site of tyrosine phosphorylation homologous to that in $pp60^{src}$. J Biol Chem 257:13877–13879
Casnellie JE, Harrison ML, Pike LJ, Hellstrom KE, Krebs EG (1982b) Phosphorylation of synthetic peptides by a tyrosine protein kinase from the particulate fraction of a lymphoma cell line. Proc Natl Acad Sci USA 79:282–286
Cheng Y-SE, Chen LB (1981) Detection of phosphotyrosine-containing 36,000 dalton protein in the framework of cells transformed with Rous sarcoma virus. Proc Natl Acad Sci USA 78:2388–2392
Clarke FM, Masters CJ (1974) On the association of glycolytic components in skeletal muscle extracts. Biochem Biophys Acta 358:193–207
Cohen P (1978) The role of cyclic AMP dependent protein kinase in the regulation of glycogen metabolism in mammalian skeletal muscle. In: Current topics in cellular regulation, vol 14. Academic Press, New York, pp 117–196
Cohen S, Carpenter G, King LR (1980) Epidermal growth factor-receptor-protein kinase interactions. J Biol Chem 255:4834–4842
Cohen S, Ushiro H, Stoschek C, Chinkers M (1982) A native 170,000 epidermal growth factor receptor-kinase complex from shed plasma membrane vesicles. J Biol Chem 257:1523–1531
Collett MS, Erikson RL (1978) Protein kinase activity associated with the avian sarcoma virus src gene product. Proc Natl Acad Sci USA 75:2021–2024

Collett MS, Erikson E, Purchio AF, Brugge JS, Erikson RL (1979) A normal cell protein similar in structure and function to the avian sarcoma virus transforming gene product. Proc Natl Acad Sci USA 76:3159–3163
Collett MS, Erikson E, Erikson RL (1980) Avian sarcoma virus transforming gene product pp60^{src} shows protein kinase activity specific for tyrosine. Nature 285:167–169
Cooper JA, Hunter T (1981a) Changes in protein phosphorylation in Rous sarcoma virus transformed chicken embryo cells. Mol Cell Biol 1:165–178
Cooper JA, Hunter T (1981b) Four different classes of retroviruses induce phosphorylation of tyrosines present in similar cellular proteins. Mol Cell Biol 1:394–407
Cooper JA, Hunter T (1981c) Similarities and differences between the effects of epidermal growth factor and Rous sarcoma virus. J Cell Biol 91:878–883
Cooper JA, Hunter T (1982) Discrete primary locations of a tyrosine protein kinase and of three proteins that contain phosphotyrosine in virally transformed chick fibroblasts. J Cell Biol 94:287–296
Cooper JA, Hunter T (1983) Identification and characterization of cellular targets for tyrosine protein kinases. J Biol Chem 258:1108–1115
Cooper JA, Bowen-Pope D, Raines E, Ross R, Hunter T (1982) Similar effects of platelet-derived growth factor and epidermal growth factor on the phosphorylation of tyrosine in cellular proteins. Cell 31:263–273
Cooper JA, Nakamura KD, Hunter T, Weber MJ (1983a) Phosphotyrosine-containing proteins and the expression of transformation parameters in cells infected with partial transformation mutants of Rous sarcoma virus. J Virol 46:15–28
Cooper JA, Reiss NA, Schwartz RJ, Hunter T (1983b) Three glycolytic enzymes are phosphorylated at tyrosine in cells transformed by Rous sarcoma virus. Nature 302:218–223
Cooper JA, Sefton BM, Hunter T (1983c) The detection of phosphotyrosine in proteins. In: Corbin JD, Hardman JG (eds) Hormone action: protein kinases. Methods in Enzymology, Vol 99 Academic Press, New York, pp 387–402
Courtneidge SA, Bishop JM (1982) Transit of pp60$^{v\text{-}src}$ to the plasma membrane. Proc Natl Acad Sci USA 79:7117–7121
Courtneidge SA, Levinson AD, Bishop JM (1980) The protein encoded by the transforming gene of avian sarcoma virus (pp60^{src}) and a homologous protein in normal cells (pp60$^{proto\text{-}src}$) are associated with the membrane. Proc Natl Acad Sci USA 77:3783–3787
Courtneidge S, Ralston R, Alitalo K, Bishop JM (1983) The subcellular location of an abundant substrate (p36) for tyrosine-specific protein kinases. Mol Cell Biol 3:340–350
Cross F, Hanafusa H (1983) Local mutagenesis of Rous sarcoma virus: the major sites of tyrosine and serine phosphorylation of p60^{src} are dispensable for transformation. Cell 34:597–608
Czernilofsky AP, Levinson AD, Varmus HE, Bishop JM, Tischler E, Goodman HM (1980) Nucleotide sequence of an avian sarcoma virus oncogene (*src*) and proposed amino acid sequence for the gene product. Nature 287:193–203
David-Pfeuty T, Singer SJ (1980) Altered distribution of the cytoskeletal proteins vinculin and α-actinin in cultured fibroblasts transformed by Rous sarcoma virus. Proc Natl Acad Sci USA 77:6687–6691
Decker S (1982) Phosphorylation of the $M_r = 34{,}000$ protein in normal and Rous sarcoma virus transformed rat fibroblasts. Biochem Biophys Res Commun 109:434–441
Eckhart W, Hutchinson MA, Hunter T (1979) An activity phosphorylating tyrosine in polyoma T antigen immunoprecipitates. Cell 18:925–933
Edelman GM, Yahara I (1976) Temperature sensitive changes in surface modulating assemblies of fibroblasts transformed by mutants of Rous sarcoma virus. Proc Natl Acad Sci USA 73:2047–2051
Ek B, Heldin C-H (1982) Characterization of a tyrosine-specific kinase activity in human fibroblast membranes stimulated by platelet-derived growth factor. J Biol Chem 257:10486–10492
Ek B, Westermark B, Wasteson A, Heldin C-H (1982) Stimulation of tyrosine-specific phosphorylation by platelet-derived growth factor. Nature 295:419–420
Erikson E, Erikson RL (1980) Identification of a cellular protein substrate phosphorylated by the avian sarcoma virus transforming gene product. Cell 21:829–836

Erikson E, Cook R, Miller GJ, Erikson RL (1981 a) The same normal cell protein is phosphorylated after transformation by avian sarcoma viruses with unrelated transforming genes. Mol Cell Biol 1:43–50

Erikson E, Shealy DJ, Erikson RL (1981 b) Evidence that viral transforming gene products and epidermal growth factor stimulate phosphorylation of the same cellular protein with similar specificity. J Biol Chem 256:11381–11384

Erikson RL, Collett MS, Erikson E, Purchio AF (1979) Evidence that the avian sarcoma virus transforming gene product is a cyclic-AMP independent protein kinase. Proc Natl Acad Sci USA 76:6260–6264

Feldman RA, Hanafusa T, Hanafusa H (1980) Characterization of protein kinase activity associated with the transforming gene product of Fujinami sarcoma virus. Cell 22:757–765

Feldman RA, Wang L-H, Hanafusa H, Balduzzi PC (1982) Avian sarcoma virus UR2 encodes a transforming protein which is associated with a unique protein kinase activity. J Virol 42:228–236

Foulkes JG, Howard RF, Ziemiecki A (1981) Detection of a novel mammalian protein phosphatase with an activity for phosphotyrosine. FEBS Lett 130:197–200

Foulkes JG, Erikson E, Erikson RL (1983) Separation of multiple phosphotyrosyl-protein phosphatases from chicken brain. J Biol Chem 258:431–438

Frackelton AR, Ross AH, Eisen HN (1983) Characterization and use of monoclonal antibodies for isolation of phosphotyrosyl proteins from retrovirus-transformed cells and growth factor-stimulated cells. Mol Cell Biol 3:1343–1352

Fulton AB (1982) How crowded is the cytoplasm? Cell 30:345–347

Geiger B (1979) A 130K protein from chicken gizzard: its location at the termini of microfilament bundles in cultured chicken cells. Cell 18:193–205

Geiger B (1982) Microheterogeneity of avian and mammalian vinculin: distinctive subcellular distribution of different isovinculins. J Mol Biol 159:685–701

Ghosh-Dastidar P, Fox CF (1983) Epidermal growth factor and epidermal growth factor receptor-dependent phosphorylation of a Mr = 34,000 protein substrate for pp60^{src}. J Biol Chem 258:2041–2044

Ghysdael J, Neil JC, Vogt PK (1981 a) A third class of avian sarcoma viruses defined by related transformation specific proteins of Yamaguchi 73 and Esh sarcoma virus. Proc Natl Acad Sci USA 78:2611–2615

Ghysdael J, Neil JC, Wallbank AM, Vogt PK (1981 b) Esh sarcoma virus codes for a *gag*-linked transformation-specific protein with an associated protein kinase activity. Virology 111:386–400

Gilmer TM, Erikson RL (1981) Rous sarcoma virus transforming protein, p60^{src}, expressed in *E. coli*, functions as a protein kinase. Nature 294:771–773

Gilmore TD, Radke K, Martin GS (1982) Tyrosine phosphorylation of a 50K cellular polypeptide associated with the Rous sarcoma virus transforming protein pp60^{src}. Mol Cell Biol 2:199–206

Greenberg ME, Edelman GM (1983 a) Comparison of the 34,000 dalton pp60^{src} substrate and a 38,000 dalton phosphoprotein identified by monoclonal antibodies. J Biol Chem 258:8497–8502

Greenberg ME, Edelman GM (1983 b) The 34kD-pp60^{src} substrate is located at the plasma membrane in normal and RSV-transformed cells. Cell 33:767–779

Hanafusa T, Mathey-Prevot B, Feldman RA, Hanafusa H (1981) Mutants of Fujinami sarcoma virus which are temperature-sensitive for cellular transformation and protein kinase activity. J Virol 38:347–355

Hampe A, Laprevotte I, Galibert F, Fedele LA, Sherr C (1982) Nucleotide sequences of feline retroviral oncogenes (*v-fes*) provide evidence for a family of tyrosine-specific protein kinase genes. Cell 30:775–785

Hirano A, Vogt PK (1981) Avian sarcoma virus PRCII: conditional mutants temperature sensitive in the maintenance of fibroblast transformation. Virology 109:193–197

Hunter T (1982 a) Synthetic peptide substrates for a tyrosine protein kinase. J Biol Chem 257:4843–4848

Hunter T (1982 b) Phosphotyrosine – a new protein modification. Trends Biochem Sci 7:246–249

Hunter T, Sefton BM (1980) The transforming gene product of Rous sarcoma virus phosphorylates tyrosine. Proc Natl Acad Sci USA 77:1311–1315

Hunter T, Cooper JA (1981) Epidermal growth factor induces rapid tyrosine phosphorylation of proteins in A431 human tumor cells. Cell 24:741–752

Hunter T, Sefton BM (1982) Protein kinases and viral transformation. In: Cohen P, Van Heyningen S (eds) The molecular actions of toxins and viruses. Elsevier/North Holland, Amsterdam, pp 333–366 (Molecular aspects of cellular regulation, vol 2)

Hunter T, Cooper JA (1983a) A comparison of the tyrosine protein kinases encoded by retroviruses and activated by growth factors. In: Bradshaw R, Gill G, Fox CF (eds) Evolution of hormone/receptor systems, UCLA symposia. Liss, New York, pp 369–382

Hunter T, Cooper JA (1983b) The role of tyrosine phosphorylation in malignant transformation and in cellular growth control. In: Cohn W (ed) Prog Nucleic Acid Res Mol Biol, vol 29. Academic, New York, pp 221–233

Ito S, Richert N, Pastan I (1982) Phospholipids stimulate phosphorylation of vinculin by the tyrosine-specific protein kinase of Rous sarcoma virus. Proc Natl Acad Sci USA 79:4628–4631

Kasuga M, Karlsson FA, Kahn CR (1981) Insulin stimulates the phosphorylation of the 95,000-dalton subunit of its own receptor. Science 215:185–187

Kawai S, Yoshida M, Segawa K, Sugiyama H, Ishizaki R, Toyoshima K (1980) Characterization of Y73, an avian sarcoma virus: A unique transforming gene and its product, a phosphopolyprotein with protein kinase activity. Proc Natl Acad Sci USA 77:6199–6203

King LE Jr, Carpenter G, Cohen S (1980) Characterization by electrophoresis of epidermal growth factor stimulated phosphorylation using A-431 membranes. Biochemistry 19:1524–1528

Kitamura N, Kitamura A, Toyoshima K, Hirayama Y, Yoshida M (1982) Avian sarcoma virus Y73 genome sequence and structural similarity of its transforming gene product to that of Rous sarcoma virus. Nature 297:205–208

Kobayashi N, Kaji A (1980) Phosphoprotein associated with activation of the *src* gene product in myogenic cells. Biochem Biophys Res Commun 93:278–284

Kobayashi N, Tanaka A, Kaji A (1981) In vitro phosphorylation of the 36K protein in extract from Rous sarcoma virus-transformed chicken fibroblasts. J Biol Chem 256:3053–3058

Krueger JG, Garber EA, Goldberg AR, Hanafusa H (1982) Changes in amino-terminal sequences of $pp60^{src}$ lead to decreased membrane association and decreased in vivo tumorigenicity. Cell 28:889–896

Lanks KW, Kasambalides EJ, Chinkers M, Brugge JS (1982) A major cytoplasmic glucose-regulated protein is associated with the Rous sarcoma virus transforming $pp60^{src}$ protein. J Biol Chem 257:8604–8607

Laszlo A, Radke K, Chin S, Bissell MJ (1981) Tumor promoters alter gene expression and protein phosphorylation in avian cells in culture. Proc Natl Acad Sci USA 78:6241–6245

Levinson AD, Oppermann H, Levintow L, Varmus HE, Bishop JM (1978) Evidence that the transforming gene of avian sarcoma virus encodes a protein kinase associated with a phosphoprotein. Cell 15:561–572

Levinson AD, Oppermann H, Varmus HE, Bishop JM (1980) The purified protein product of the transforming gene of avian sarcoma virus phosphorylates tyrosine. J Biol Chem 255:11973–11980

Levinson AD, Courtneidge SA, Bishop JM (1981) Structural and functional domains of the Rous sarcoma virus transforming protein ($pp60^{src}$). Proc Natl Acad Sci USA 78:1624–1628

Lipsich LA, Cutt JR, Brugge JS (1982) Association of the transforming proteins of Rous, Fujinami, and Y73 avian sarcoma viruses with the same two cellular proteins. Mol Cell Biol 2:875–880

Manai M, Cozzone AJ (1982) Two-dimensional separation of phosphoamino acids from nucleoside monophosphates. Anal Biochem 124:12–18

Maness PF, Engeser H, Greenberg ME, O'Farrell M, Gall WE, Edelman GM (1979) Characterization of the protein kinase activity of avian sarcoma virus src gene product. Proc Natl Acad Sci USA 76:5028–5032

Martinez R, Nakamura KD, Weber MJ (1982) Identification of phosphotyrosine-containing proteins in untransformed and Rous sarcoma virus-transformed chicken embryo cells. Mol Cell Biol 2:653–665

Mathey-Prevot B, Hanafusa H, Kawai S (1982) A cellular protein is immunologically cross-reactive with and functionally homologous to the Fujinami sarcoma virus transforming protein. Cell 28:897–906

McGrath JP, Levinson AD (1982) Bacterial expression of an enzymatically active protein encoded by RSV *src* gene. Nature 295:423–425

Nakamura KD, Weber MJ (1982) Phosphorylation of a 36,000 M_r cellular protein in cells infected with partial transformation mutants of Rous sarcoma virus. Mol Cell Biol 2:147–153

Nakamura KD, Martinez R, Weber MJ (1983) Tyrosine phosphorylation of specific proteins following mitogen stimulation of chicken embryo fibroblasts. Mol Cell Biol 3:380–389

Neil JC, Delamarter JF, Vogt PK (1981a) Evidence for three classes of avian sarcoma viruses: comparison of the transformation-specific gene products of PRC II, Y73, and Fujinami viruses. Proc Natl Acad Sci USA 78:1906–1911

Neil JC, Ghysdael J, Vogt PK (1981b) Tyrosine-specific protein kinase activity associated with p105 of avian sarcoma virus PRCII. Virology 109:223–228

Neil JC, Ghysdael J, Vogt PK, Smart JE (1981c) Homologous tyrosine phosphorylation sites in transformation-specific gene products of distinct avian sarcoma viruses. Nature 291:675–677

Nigg EA, Sefton BM, Hunter T, Walter G, Singer SJ (1982) Immunofluorescent localization of the transforming protein of Rous sarcoma virus with antibodies against a synthetic *src* peptide. Proc Natl Acad Sci USA 79:5322–5326

Nigg EA, Cooper JA, Hunter T (1983) Immunofluorescent localization of a 39,000 dalton substrate of tyrosine protein kinases to the cytoplasmic surface of the plasma membrane. J Cell Biol 96:1601–1609

Nishimura J, Huang JS, Deuel TF (1982) Platelet-derived growth factor stimulates tyrosine-specific protein kinase activity in Swiss mouse 3T3 cell membranes. Proc Natl Acad Sci USA 79:4303–4307

Oppermann H, Levinson AD, Levintow L, Varmus HE, Bishop JM (1979) Uninfected vertebrate cells contain a protein that is closely related to the product of the avian sarcoma virus transforming gene (src). Proc Natl Acad Sci USA 76:1804–1808

Oppermann H, Levinson W, Bishop JM (1981a) A cellular protein that associates with the transforming protein of Rous sarcoma virus is also a heat-shock protein. Proc Natl Acad Sci USA 78:1067–1071

Oppermann H, Levinson AD, Levintow L, Varmus HE, Bishop JM, Kawai S (1981b) Two cellular proteins that immunoprecipitate with the transforming protein of Rous sarcoma virus. Virology 113:736–751

Patschinsky T, Sefton BM (1981) Evidence that there exist four classes of RNA tumor viruses which encode proteins with associated tyrosine protein kinase activities. J Virol 39:104–114

Patschinsky T, Hunter T, Esch FS, Cooper JA, Sefton BM (1982) Analysis of the sequence of amino acids surrounding sites of tyrosine phosphorylation. Proc Natl Acad Sci USA 79:973–977

Pawson T, Guyden J, Kung T-H, Radke K, Gilmore T, Martin GS (1980) A strain of Fujinami sarcoma virus which is temperature-sensitive in protein phosphorylation and cellular transformation. Cell 22:767–776

Penman S, Fulton A, Capco D, Ben Ze'ev A, Wittelsberger S, Tse CF (1981) Cytoplasmic and nuclear architecture in cells and tissue: form, functions and mode of assembly. Cold Spring Harbor Symp Quant Biol 46:1013–1028

Petruzzelli LM, Ganguly S, Smith CJ, Cobb MH, Rubin CH, Rosen OM (1982) Insulin activates a tyrosine-specific protein kinase in extracts of 3T3-L1 adipocytes and human placenta. Proc Natl Acad Sci USA 79:6792–6796

Pike LJ, Gallis B, Casnellie JE, Bornstein P, Krebs EG (1982) Epidermal growth factor stimulates the phosphorylation, of synthetic tyrosine-containing peptides, by A431-cell membranes. Proc Natl Acad Sci USA 79:1443–1447

Ponticelli AS, Whitlock CA, Rosenberg N, Witte ON (1982) In vivo tyrosine phosphorylations of the Abelson virus transforming protein are absent in its normal cellular homolog. Cell 29:953–960
Presek P, Glossmann H, Eigenbrodt E, Schoner W, Rubsamen H, Friis RR, Bauer H (1980) Similarities between a phosphoprotein ($pp60^{src}$)-associated protein kinase of Rous sarcoma virus and a cyclic adenosine 3′:5′-monophosphate-independent protein kinase that phosphorylates pyruvate kinase type M_2. Cancer Res 40:1733–1741
Racker E (1972) Bioenergetics and the problem of tumor growth. Am Sci 60:56–63
Radke K, Martin GS (1979) Transformation by Rous sarcoma virus: Effects of *src* gene expression on the synthesis and phosphorylation of cellular polypeptides. Proc Natl Acad Sci USA 76:5212–5216
Radke K, Gilmore T, Martin GS (1980) Transformation by Rous sarcoma virus: a cellular substrate for transformation-specific protein phosphorylation contains phosphotyrosine. Cell 21:821–828
Radke K, Carter VC, Moss P, Dehazya P, Schliwa M, Martin GS (1983) Membrane association of a 36,000 dalton substrate for tyrosine phosphorylation in chicken embryo fibroblasts transformed by avian sarcoma viruses. J Cell Biol 97:1601–1611
Rasheed S, Barbacid M, Aaronson S, Gardner MB (1982) Origin and properties of a new feline sarcoma virus. Virology 117:238–244
Reddy EP, Smith MJ, Srinivasan A (1983) Nucleotide sequence of Abelson murine leukemia virus genome: structural similarity of its transforming gene to other *onc* gene products with tyrosine-specific kinase activity. Proc Natl Acad Sci USA 80:3623–3627
Reynolds FH Jr, Van de Ven WJM, Stephenson JR (1980a) Abelson murine leukemia virus transformation-defective mutants with impaired P120-associated protein kinase activity. J Virol 36:374–386
Reynolds FH Jr, Van de Ven WJM, Stephenson JR (1980b) Feline sarcoma virus P115 associated protein kinase phosphorylates tyrosine. Identification of a cellular substrate conserved during evolution. J Biol Chem 255:11040–11047
Reynolds FH, Todaro GJ, Fryling C, Stephenson JR (1981a) Human transforming growth factors induce tyrosine phosphorylation of EGF receptors. Nature 292:259–262
Reynolds FH Jr, Van de Ven WJM, Blomberg J, Stephenson JR (1981b) Differences in mechanisms of transformation by independent feline sarcoma virus isolates. J Virol 38:1084–1089
Reynolds FH Jr, Oroszlan S, Blomberg J, Stephenson JR (1982a) Tyrosine phosphorylation sites common to transforming proteins encoded by Gardner and Snyder-Theilen FeSV. Virology 122:134–146
Reynolds FH Jr, Oroszlan S, Stephenson JR (1982b) Abelson murine leukemia virus P120: Identification and characterization of tyrosine phosphorylation sites. J Virol 44:1097–1101
Richert ND, Davies PJA, Jay G, Pastan IH (1979) Characterization of an immune complex kinase in immunoprecipitates of avian sarcoma virus-transformed fibroblasts. J Virol 31:695–706
Richert ND, Blithe DL, Pastan IH (1982) Properties of the *src* kinase purified from Rous sarcoma virus-induced rat tumors. J Biol Chem 257:7143–7150
Rohrschneider LR (1980) Adhesion plaques of Rous sarcoma virus transformed cells contain the *src* gene product. Proc Natl Acad Sci USA 77:3514–3518
Rohrschneider LR, Rosok M (1983) Transformation parameters and $pp60^{src}$ localization in cells infected with partial transformation mutants of Rous sarcoma virus. Mol Cell Biol 3:731–746
Rohrschneider LR, Eisenman RN, Leitch CR (1979) Identification of a Rous sarcoma virus transformation-related protein in normal avian and mammalian cells. Proc Natl Acad Sci USA 76:4479–4483
Rohrschneider LR, Rosok M, Shriver K (1981) Mechanism of transformation by Rous sarcoma virus: events within adhesion plaques. Cold Spring Harbor Symp Quant Biol 46:953–965
Rosenberg NE (1982) Abelson murine leukemia virus. In: Current topics in microbiology and immunology, vol 101. Springer, Berlin, Heidelberg, New York, pp 95–126
Rosenberg NE, Clark DR, Witte ON (1981) Abelson murine leukemia virus mutants deficient in kinase activity and lymphoid cell transformation. J Virol 36:766–774

Rosok M, Rohrschneider LR (1983) Increased phosphorylation of vinculin on tyrosine does not occur during the release of stress fibers before mitosis in normal cells. Mol Cell Biol 3:475–479

Ross AH, Baltimore D, Eisen H (1981) Phosphotyrosine-containing proteins isolated by affinity chromatography with antibodies to a synthetic hapten. Nature 294:654–656

Rubsamen H, Friis RR, Bauer H (1979) *src* gene product from different strains of avian sarcoma virus: kinetics and possible mechanism of heat inactivation of protein kinase activity from cells infected by transformation-defective, temperature-sensitive, and wild-type virus. Proc Natl Acad Sci USA 76:967–971

Rubsamen H, Saltenberger K, Friis RR, Eigenbrodt E (1982) Cytosolic malic dehydrogenase activity is associated with a putative substrate for the transforming gene product of Rous sarcoma virus. Proc Natl Acad Sci USA 79:228–232

Schwartz D, Tizard R, Gilbert W (1982) The complete nucleotide sequence of the Pr-C strain of Rous sarcoma virus. In: Weiss R, Teich N, Varmus H, Coffin J (eds) RNA tumor viruses. Cold Spring Harbor, New York, pp 1338–1356

Sefton BM, Beemon K, Hunter T (1978) Comparison of the expression of the *src* gene of Rous sarcoma virus in vitro and in vivo. J Virol 28:957–971

Sefton BM, Hunter T, Beemon K (1979) Product of in vitro translation of the Rous sarcoma virus *src* gene has protein kinase activity. J Virol 30:311–318

Sefton BM, Hunter T, Beemon K (1980a) Temperature-sensitive transformation by Rous sarcoma virus and temperature-sensitive protein kinase activity. J Virol 33:220–229

Sefton BM, Hunter T, Beemon K, Eckhart W (1980b) Phosphorylation of tyrosine is essential for cellular transformation by Rous sarcoma virus. Cell 20:807–816

Sefton BM, Hunter T, Ball EH, Singer SJ (1981a) Vinculin: a cytoskeletal substrate of the transforming protein of Rous sarcoma virus. Cell 24:165–174

Sefton BM, Hunter T, Raschke WC (1981b) Evidence that the Abelson virus protein functions in vivo as a protein kinase which phosphorylates tyrosine. Proc Natl Acad Sci USA 78:1552–1556

Sefton BM, Hunter T, Nigg E, Singer SJ, Walter G (1981c) Cytoskeletal targets for viral transforming proteins with tyrosine protein kinase activity. Cold Spring Harbor Symp Quant Biol 46:939–951

Sefton BM, Patschinsky T, Berdot C, Hunter T, Elliot T (1982) Phosphorylation and metabolism of the transforming protein of Rous sarcoma virus. J Virol 41:813–820

Sefton BM, Hunter T, Cooper JA (1983) Some lymphoid cell lines transformed by Abelson murine leukemia virus lack a major 36,000-dalton tyrosine protein kinase substrate. Mol Cell Biol 3:56–63

Shibuya M, Hanafusa H (1982) Nucleotide sequence of Fujinami sarcoma virus: evolutionary relationship of its transforming gene with transforming genes of other sarcoma viruses. Cell 30:787–795

Shibuya M, Hanafusa T, Hanafusa H, Stephenson JR (1980) Homology exists among the transforming sequences of avian and feline sarcoma viruses. Proc Natl Acad Sci USA 77:6536–6540

Shilo B-Z, Weinberg RA (1981) DNA sequences homologous to vertebrate oncogenes are conserved in *Drosophila melanogaster*. Proc Natl Acad Sci USA 78:6789–6792

Shriver K, Rohrschneider LR (1981) Organization of $pp60^{src}$ and selected cytoskeletal proteins within adhesion plaques and junctions of Rous sarcoma virus-transformed rat cells. J Cell Biol 89:525–535

Sigel P, Pette D (1969) Intracellular localization of glycogenolytic and glycolytic enzymes in white and red rabbit skeletal muscle. J Histochem Cytochem 17:225–237

Singh VN, Singh M, August JT, Horecker BL (1974) Alterations in glucose metabolism in chick-embryo cells transformed by Rous sarcoma virus: Intracellular levels of glycolytic intermediates. Proc Natl Acad Sci USA 71:4129–4132

Smart JE, Oppermann H, Czernilofsky AP, Purchio AF, Erikson RL, Bishop JM (1981) Characterization of sites for tyrosine phosphorylation in the transforming protein of Rous sarcoma virus ($pp60^{src}$) and its normal cellular homologue ($pp60^{c\text{-}src}$). Proc Natl Acad Sci USA 78:6013–6017

Snyder MA, Bishop JM, Colby WW, Levinson AD (1983) Phosphorylation of tyrosine-416 is not required for the transforming properties of and kinase activity of $pp60^{v\text{-}src}$. Cell 32:891–901
Takeya T, Feldman RA, Hanafusa A (1982) DNA sequence of the viral and cellular *src* gene of chickens. I. Complete nucleotide sequence of an EcoRI fragment of recovered avian sarcoma virus which codes for gp37 and $pp60^{src}$. J Virol 44:1–11
Ushiro H, Cohen S (1980) Identification of phosphotyrosine as a product of epidermal growth factor-activated protein kinase in A-431 cell membranes. J Biol Chem 255:8363–8365
Van de Ven WJM, Reynolds FH Jr, Stephenson JR (1980) The nonstructural components of polyproteins encoded by replication-defective mammalian transforming retroviruses are phosphorylated and have associated protein kinase activity. Virology 101:185–187
Wang E, Goldberg AR (1976) Changes in microfilament organization and surface topography upon transformation of chick embryo fibroblasts with Rous sarcoma virus. Proc Natl Acad Sci USA 73:4065–4069
Wang JYJ, Queen C, Baltimore D (1982) Expression of an Abelson murine leukemia virus-encoded protein in *Escherichia coli* causes extensive phosphorylation of tyrosine residues. J Biol Chem 257:13181–13184
Wang L-H, Feldman R, Shibuya M, Hanafusa H, Notter MFP, Balduzzi PC (1981) Genetic structure, transforming sequence and gene product of avian sarcoma virus, UR1. J Virol 40:258–267
Warburg O (1930) The metabolism of tumors. Dickens E (trans). Constable, London
Weber MJ (1973) Hexose transport in normal and in Rous sarcoma virus-transformed cells. J Biol Chem 248:2978–2983
Weber MJ, Friis RR (1979) Dissociation of transformation parameters using temperature-conditional mutants of Rous sarcoma virus. Cell 16:25–32
Willingham MC, Jay G, Pastan I (1979) Localization of the ASV *src* gene product to the plasma membrane of transformed cells by immunoelectron microscopy. Cell 18:125–134
Witte ON, Rosenberg N, Baltimore D (1979) A normal cell protein cross-reactive to the major Abelson murine leukemia virus gene product. Nature 281:396–398
Witte ON, Dasgupta A, Baltimore D (1980a) Abelson murine leukemia virus protein is phosphorylated in vitro to form phosphotyrosine. Nature 283:826–831
Witte ON, Goff S, Rosenberg N, Baltimore D (1980b) A transformation-defective mutant of Abelson murine leukemia virus lacks protein kinase activity. Proc Natl Acad Sci USA 77:4993–4997
Witte ON, Ponticelli A, Gifford A, Baltimore D, Rosenberg N, Elder J (1981) Phosphorylation of the Abelson murine leukemia virus transforming protein. J Virol 39:870–878
Wong TW, Goldberg AR (1983) In vitro phosphorylation of angiotensin analogues by tyrosyl protein kinases. J Biol Chem 258:1022–1025

Phosphotyrosyl-Protein Phosphatases

J. GORDON FOULKES

1 Introduction . . . 163
1.1 Retroviral Oncogene Products . . . 163
1.2 Tyrosine Phosphorylation . . . 164
1.3 Tyrosyl Phosphorylation and Cell Growth . . . 164
1.4 The Role of Tyrosyl-Protein Phosphatases in Cell Transformation . . . 165
2 Protein Phosphatases . . . 166
2.1 Phosphoseryl- and Phosphothreonyl-Protein Phosphatases . . . 166
2.2 Phosphotyrosyl-Protein Phosphatases . . . 167
2.2.1 Phosphotyrosyl-Protein Phosphatase Activity in Extracts – The Effects of Zn^{2+} and VO_4^{3-} . . . 167
2.2.2 The Relation of Phosphotyrosyl-Protein Phosphatases to Phosphoseryl-Protein Phosphatases . . . 168
2.2.3 The Relation of Phosphotyrosyl-Protein Phosphatases to Acid and Alkaline Phosphatases . . . 172
3 Phosphotyrosine Phosphatases . . . 173
4 Conclusions . . . 174
References . . . 176

1 Introduction

1.1 Retroviral Oncogene Products

Extensive genetic evidence indicates that the protein product of a single gene of RNA tumor viruses is responsible for the transformation of virus-infected cells to the malignant state (HANAFUSA 1977; DUESBERG and BISTER 1981; BISHOP and VARMUS 1982; LINIAL and BLAIR 1982; VARMUS 1982).

Retroviral transforming genes, termed *onc* genes, appear to have been derived from host cell sequences. A wide range of metazoan species, from *Drosophila melanogaster* to *Homo sapiens,* contain DNA which is complementary to the viral *onc* genes (SPECTOR et al. 1978; ERIKSON et al. 1980; DUESBERG and BISTER 1981; EVA et al. 1982; HOFFMAN-FALK et al. 1983). These regions of DNA homology are expressed in normal cells, which has led to the view that their protein products may have a role in the control of normal cell growth and differentiation (COLLETT et al. 1979; ERIKSON

Center for Cancer Research, Massachusetts Institute of Technology, Cambridge, MA 02139, USA

Current Topics in Microbiology and Immunology, Vol. 107

et al. 1980; BISHOP and VARMUS 1982; MULLER et al. 1982; SCHARTL and BARNEKOW 1982; GOYETTE et al. 1983).

Over the past few years, originating primarily with studies described in Erikson's laboratory (COLLETT et al. 1979; ERIKSON et al. 1980), the protein products of a number of different *onc* genes have been identified and found to be associated with a protein kinase activity (ERIKSON et al. 1980; BEEMON 1981; HUNTER and SEFTON 1981; BISHOP 1982; BISHOP and VARMUS 1982; COOPER and HUNTER 1983).

1.2 Tyrosine Phosphorylation

Since the discovery of the cyclic AMP-dependent protein kinase in 1968 (WALSH et al. 1968), reversible phosphorylation of proteins has come to be accepted as the major mechanism for the regulation of protein function in eukaryotic systems (NIMMO and COHEN 1977; KREBS and BEAVO 1979; WELLER 1979; COHEN 1980; COHEN et al. 1981; COHEN 1982; COOPER et al. 1982).

Several amino acid phosphate acceptors have been identified, with phosphoserine being the most abundant phosphoamino acid (TABORSKY 1974; WELLER 1979; HUNTER and SEFTON 1981).

Phosphotyrosine was originally described as a free phosphoamino acid found at high concentrations in *Drosophila* larvae (MITCHELL and LUNAN 1964), and later in the adenylylated form of glutamine synthetase in *Escherichia coli* (SHAPIRO and STADTMAN 1968) and in the protein-RNA linkage of the poliovirus genome (AMBROS and BALTIMORE 1978; ROTHBERG et al. 1978). In vitro phosphorylation of proteins on tyrosine residues was first demonstrated in immunoprecipitates containing either polyoma middle T antigen (ECKHART et al. 1979) or the transforming protein of Abelson murine leukemia virus (WITTE et al. 1980).

1.3 Tyrosyl Phosphorylation and Cell Growth

During the last few years, the transforming gene protein products of many RNA tumor viruses have been shown to be associated with tyrosine-specific protein kinases. These enzymes account for a five- to tenfold increase in the phosphotyrosine content of cellular proteins following viral-induced transformation of cells by these viruses (HUNTER and SEFTON 1981). However, the function of most of these phosphotyrosyl proteins remains to be established (RADKE and MARTIN 1979; HUNTER and SEFTON 1981; COOPER and HUNTER 1981; COOPER et al. 1982; COOPER and HUNTER 1983; COOPER et al. 1983).

The current model for RNA tumor virus-induced cell transformation depicts the phosphorylation of certain cellular proteins by a tyrosine-specific protein kinase, either directly or indirectly through a kinase–kinase cascade (ERIKSON et al. 1980; HUNTER and SEFTON 1981; BISHOP 1982; BISHOP and

VARMUS 1982). Modulation of host cell protein function, via modifications such as phosphorylation, could account for the multitude of altered cell parameters which are characteristic of the neoplastic phenotype (HANAFUSA 1977). Since normal uninfected cells contain low levels of homologous tyrosyl-protein kinases (ERIKSON et al. 1980; HUNTER and SEFTON 1981; SCHARTL and BARNEKOW 1982), transformation has been visualized as a process which is the result of either (a) an increase in total tyrosine kinase activity, which enhances the degree of phosphorylation of normal cell protein substrates, or (b) the introduction into normal cells of a mutated viral *onc* gene product, with an altered substrate specificity. This could result in the phosphorylation of proteins, or sites, which are not normally phosphorylated by the homologous cellular kinase. Mutations in the viral *onc* gene could also affect the regulation or subcellular localization of the kinase.

The idea that tyrosyl-protein kinases may also be involved in the regulation of normal cell growth receives support from several lines of evidence, including (a) the observation that tyrosyl-protein kinases exist in a broad spectrum of normal cells; (b) the finding that transcription of these homologous cellular genes varies during embryogenesis (MULLER et al. 1982); and (c) the finding that the receptors for epidermal growth factor, platelet-derived growth factor, and insulin are all associated with tyrosine-specific protein kinases (USHIRO and COHEN 1980; REYNOLDS et al. 1981; AVRUCH et al. 1982; BUHROW et al. 1982; COOPER et al. 1982; EK and HELDIN 1982; EK et al. 1982; KASUGA et al. 1982a, b; ZICK et al. 1983). However, as in the case of the tyrosyl-kinases associated with cell transformation, the role of similar activities in normal cells is unresolved, since the physiological substrates for these enzymes have not been identified.

1.4 The Role of Tyrosyl-Protein Phosphatases in Cell Transformation

In all systems which are regulated by phosphorylation, proteins are phosphorylated in a reversible manner. Therefore, protein phosphatases should exist which are capable of reversing the actions of the tyrosyl-protein kinases. The existence of phosphotyrosyl-protein phosphatases was first suggested by work involving tumor viruses with temperature-sensitive (ts) transforming proteins. Cells infected by either Rous sarcoma virus or Fujinami sarcoma virus with *ts onc* genes can undergo temperature-dependent reversible transformation, even in the presence of inhibitors of protein synthesis (FRIIS et al. 1979; HUNTER et al. 1979; RADKE and MARTIN 1979; ZIEMIECKI and FRIIS 1980; HUNTER and SEFTON 1981; LEE W. et al. 1981). In several cases, this has been correlated with a similar temperature-sensitive activity of the corresponding viral protein kinase in vitro (HUNTER et al. 1979; ERIKSON et al. 1980; ZIEMIECKI and FRIIS 1980; BEEMON 1981; HUNTER and SEFTON 1981). In cells infected with such *ts* mutants, phosphotyrosine levels decrease to those observed in normal cells within 1–2 h following a shift to the nonpermissive temperature (RADKE and MARTIN 1979; SEFTON et al. 1980; COOPER and HUNTER 1981). These effects appear to be independent of protein turnover. Clearly the inactivation of a temperature-sensitive protein kinase in vivo would have no effect on the phosphorylation state of a target protein, unless a phosphatase was present to reverse the actions of the viral kinase. Such observations demonstrate the potential importance of phosphotyrosyl-protein phosphatases in controlling the transformation state. Models for virally induced cell transformation should include these activities. Thus the transformation event should be envisaged as a dynamic kinase phosphatase equilibrium process, rather than a change in kinase activity per se.

2 Protein Phosphatases

2.1 Phosphoseryl- and Phosphothreonyl-Protein Phosphatases

Following the discovery of the cyclic AMP-dependent protein kinase as a major intracellular mediator of hormone action, many laboratories focused their attention on phosphorylation events. Dephosphorylation received less attention and was often considered to be an unregulated process. However, this viewpoint has changed dramatically in recent years, with the demonstration that protein phosphatases are regulated by a variety of hormones and are controlled by such factors as Ca^{2+}, calmodulin, and cyclic AMP (COHEN 1978; CURNOW and LARNER 1979; FOULKES and COHEN 1979; COHEN 1980; GREEN et al. 1980; COHEN et al. 1981; COHEN 1982; FOULKES et al. 1982; STEWART et al. 1982; STEWART et al. 1983).

Recently, vertebrate phosphoseryl- and phosphothreonyl-protein phosphatases, which have been identified in a wide range of tissues and species, have been classified into two major groups. This classification is based on substrate specificity, sensitivity toward two heat stable inhibitor proteins, and subunit composition of the purified enzymes (COHEN et al. 1981; STEWART et al. 1981; COHEN 1982; FOULKES and MALLER 1982; FOULKES et al. 1983a, b; INGEBRITSEN et al. 1983a; PATO et al. 1983; STEWART et al. 1983).

Each protein phosphatase group appears to contain several distinct molecular species. Two forms of the type 1 protein phosphatases and three forms of the type 2 enzymes have been described. One major distinction between type 1 and type 2 protein phosphatases is that type 1 phosphatases (designated PP1), but not type 2 phosphatases (PP2), are inhibited by two thermostable proteins, termed inhibitor-1 and inhibitor-2 (NIMMO and COHEN 1978a, b; FOULKES and COHEN 1979, 1980; COHEN et al. 1981; FOULKES et al. 1982). The specificity of these inhibitors has allowed these proteins to be used as highly selective probes to define the function of the type 1 enzymes in a variety of systems. The catalytic subunit of protein phosphatase 1 is a spontaneously active enzyme requiring no activation. A species of PP1 ($M_r = 35000$) has been purified (INGEBRITSEN et al. 1980; LEE et al. 1980; LEE EYC et al. 1981), but the structure of the native enzyme has not been clarified; species as large as $M_r = 250000$ have been reported (Co-

HEN 1978; LEE et al. 1980; LEE EYC et al. 1981; LI 1982). A second form of PP1 exists which requires MgATP and at least three polypeptides, designated F_A, F_c, and inhibitor-2, for activity (VANDENHEEDE et al. 1980; Yang et al. 1980; COHEN et al 1981; Hemmings et al. 1981; STEWART et al. 1981; VANDENHEEDE et al. 1981a, b; INGEBRITSEN et al. 1983a). The phosphatase activity resides in the F_c component. The MgATP dependency appears to involve phosphorylation of inhibitor-2 by F_A, which inactivates inhibitor-2, thus allowing the phosphatase activity to be expressed (HEMMINGS et al. 1982; RESINK et al. 1983). Type 1 phosphatases are regulated in vivo by both epinephrine and insulin (FOULKES and COHEN 1979; FOULKES et al. 1982).

Type 2 protein phosphatases include three enzymes termed PP2A, PP2B, and PP2C. Type 2A protein phosphatases exist in several different forms. $PP2A_1$ contains three different subunits (M_r=60000, 55000, and 38000), while $PP2A_2$ contains just two of these (M_r=60000 and 38000) (CROUCH and SAFER 1980; TAMURA and TSUIKI 1980; TAMURA et al. 1980; COHEN et al. 1981; PATO et al. 1983). The M_r=38000 entity of $PP2A_1$ and $PP2A_2$ is the catalytic subunit of these species and is distinct from the M_r=35000 polypeptide of the type 1 phosphatases (COHEN et al. 1980; INGEBRITSEN et al. 1980; LEE et al. 1980; LEE EYC et al. 1981; INGEBRITSEN et al. 1983a). PP2B contains two subunits (M_r=61000 and 19000) and is dependent on Ca^{2+}-calmodulin for optimal activity (STEWART et al. 1982; STEWART et al. 1983). The M_r=15000 subunit binds Ca^{2+}, while the M_r=61000 subunit interacts with calmodulin and may represent the catalytic domain. PP2C has a single subunit (M_r=46000) (COHEN et al. 1981; LI 1981, 1982; INGEBRITSEN et al. 1983). This enzyme was originally described as a phosphoseryl-casein phosphatase, with optimal activity in the presence of Mg^{2+} (LI et al. 1978). This enzyme has also been purified by PATO and ADELSTEIN (1980) as a myosin light chain phosphatase. The tissue distribution, substrate specificities, and metabolic roles of the type 1 and type 2 protein phosphatases are discussed elsewhere (COHEN 1980; CROUCH and SAFER 1980; COHEN et al. 1981; COHEN 1982; FOULKES and MALLER 1982; FOULKES et al. 1983a and 1983b; INGEBRITSEN et al. 1983b; STEWART et al. 1983).

2.2 Phosphotyrosyl-Protein Phosphatases

2.2.1 Phosphotyrosyl-Protein Phosphatase Activity in Extracts – The Effects of Zn^{2+} and VO_4^{3-}

Dephosphorylation of phosphotyrosyl-proteins in vitro was first observed by COHEN and his colleagues, in membrane preparations from the human epidermoid carcinoma cell line A431 (CARPENTER et al. 1979). Addition of epidermal growth factor (EGF) to these membranes stimulated phosphorylation of the EGF receptor on tyrosine residues[1] (CARPENTER et al. 1979;

1 At the time of publication, the phosphorylated amino acid was incorrectly reported to be phosphothreonine instead of phosphotyrosine (USHIRO and COHEN 1980)

than phosphoseryl-casein. In particular, PPT-2 was 1000 times more active in dephosphorylating phosphotyrosyl-casein.

Since EDTA-fluoride failed to inhibit these enzymes, we examined other putative inhibitors, and found that 30 m*M* pyrophosphate inhibited all three activities by greater than 95%. We also noted that 1 m*M* Mn^{2+}, but not Mg^{2+}, significantly inhibited these activities. This suggested that the apparent preference of certain tyrosyl-protein kinases for MnATP over MgATP (Hunter and Sefton 1981) could be mediated by an inhibition of phosphatase activity, rather than a change in the kinase activity per se.

Recently, Horlein et al. (1982) also have separated multiple phosphotyrosyl-protein phosphatases, using extracts from Ehrlich ascites tumor cells. The principal substrate in these studies was a chemically modified glycogen phosphorylase, phosphorylated in vitro by the EGF receptor kinase from A431 cells. When cell lysates were fractionated on DEAE-Sephadex, 60% of the phosphotyrosyl-protein phosphatase was found in the flow-through fraction. The activity which bound to the column separated into two peaks. After elution, the major activity (30%–35%) was further fractionated on Zn^{2+}-agarose chromatography and again separated into two peaks. The major activity at this step had an M_r of 40000 on gel filtration and was essentially free of phosphoseryl-glycogen phosphatase activity. Micromolar levels of Zn^{2+} were found to inhibit the $M_r=40000$ enzyme, but the degree of inhibition was dependent on the substrate employed. The dephosphorylation of phosphotyrosyl-membrane proteins from A431 cells was inhibited 50% by 5 μ*M* Zn^{2+}, while 33 μ*M* Zn^{2+} was required to inhibit the dephosphorylation of phosphotyrosyl-phosphorylase by 50%.

In our studies using the chicken brain extract (Foulkes et al. 1983a), approximately 10% of the phosphotyrosyl-casein phosphatase activity fractionated with the phosphoseryl-protein phosphatases $PP2A_1$ and $PP2A_2$, and showed optimal activity in the presence of Mn^{2+}. Chicken brain $PP2A_1$ and $PP2A_2$ migrated on gel filtration with apparent M_r of 250000 and 110000 respectively[1]. However, when these enzymes were treated with ethanol to release the catalytic subunit ($M_r=35000$) (Lee et al. 1980), both phosphotyrosyl- and phosphoseryl-protein phosphatase activities were again found to co-migrate, but now with an M_r of 35000. Furthermore, the relative phosphotyrosyl- and phosphoseryl-protein phosphatase activities of $PP2A_1$, $PP2A_2$, and the $M_r=35000$ catalytic subunit were identical. These observations strongly suggested that the type 2A enzymes have an inherent phosphotyrosyl-protein phosphatase activity. Data supporting this have also been obtained by Li and co-workers (Li 1982; Chernoff et al 1983a). In these studies essentially homogeneous preparations of bovine cardiac $PP2A_2$[2] were found to dephosphorylate both phosphotyrosyl-IgG and

1 The molecular weight of $PP2A_2$ has been variously reported between 80000 and 150000 (Tamura et al. 1980; Crouch and Safer 1980; Cohen et al. 1981; Foulkes et al. 1983a; Ingebritsen et al. 1983). Whether this reflects tissue variation, the presence of tetrameric and dimeric forms, or both, remains to be determined

2 Dr. Li's laboratory has previously designated $PP2A_2$ and PP2C as type 1 and type 2 respectively. This reflects differences in nomenclature only, and not enzyme type. For the sake of clarity, the former nomenclature has been retained throughout the text

BRAUTIGAN et al. 1982; LEIS and KAPLAN 1982; SWARUP et al. 1982a, b). In our own early studies, we described a phosphotyrosyl-protein phosphatase activity in extracts from normal rat liver and skeletal muscle which, in contrast to all known protein phosphatases, was insensitive to either EDTA or EDTA (4 m*M*)-fluoride (50 m*M*) (FOULKES et al. 1981). The substrate used in these studies was phosphotyrosyl-IgG, which had been labeled in vitro by the Rous sarcoma virus tyrosyl-protein kinase, $pp60^{src}$. This work indicated that this phosphotyrosyl-protein phosphatase activity was distinct from all protein phosphatases described previously. Although other reports also suggested that fluoride did not inhibit phosphotyrosyl-protein phosphatase activity, it has been demonstrated that it is essential to carry out this reaction in the presence of EDTA[1], since contamination by low levels of divalent cations can overcome fluoride inhibition. For example, 1 m*M* Mn^{2+} can reverse inhibition by 50 m*M* fluoride (ANTONIW and COHEN 1976; INGEBRITSEN et al. 1983b).

In our subsequent studies on the phosphotyrosyl-protein phosphatase present in normal tissues, we fractionated extracts from chicken brain (which contains high levels of cellular $pp60^{src}$; SCHARTL and BARNEKOW 1982)[2], using ^{32}P-phosphotyrosyl-casein, labeled by $pp60^{src}$, as a substrate (FOULKES et al. 1983a). As observed with rat tissue extracts (FOULKES et al. 1981), the chicken brain phosphotyrosyl-protein phosphatase activity was insensitive to EDTA-fluoride[3], while the phosphoseryl-protein phosphatase activity was completely inhibited.

Following fractionation of the chicken brain extract on DEAE-cellulose, three major peaks of phosphotyrosyl-protein phosphatase activity were observed and designated PPT-1, PPT-2, and PPT-3 in order of elution. PPT-1, which did not bind to the column at pH 7.0, constituted 18% of the total activity. By contrast, only 0.4% of the phosphoseryl-protein phosphatase activity was found in this fraction. PPT-1 displayed a polydisperse activity profile on gel filtration, with an M_r range of 30000 to 100000. PPT-2, the major activity (56%), eluted at 145 m*M* NaCl and displayed an M_r of 43000 on gel filtration. PPT-3, 16% of the total activity, eluted at 230 m*M* NaCl and had an M_r of 95000. All three enzymes showed optimal activity in the presence of 1 m*M* EDTA and all three activities were insensitive to 1 m*M* EDTA–50 m*M* fluoride. Following gel filtration, the three preparations were considerably more active toward phosphotyrosyl-casein

1 FUKAMI and LIPMANN (1982) have recently reported a phosphatase activity, isolated from *Drosophila* larvae, which dephosphorylated free phosphotyrosine. This activity was inhibited by fluoride, but not by EDTA-fluoride. No data were presented to suggest that this enzyme might also function as a protein phosphatase

2 Protein phosphatase activity is routinely determined by measuring ^{32}P released from a ^{32}P-labeled phosphoprotein into a trichloroacetic acid soluble fraction. However, it is important to distinguish between this activity and ^{32}P released as phosphopeptides by proteolysis. This may be achieved by separation of inorganic ^{32}P into an organic phase by formation of the phosphomolybdate complex

3 It is important to compare the effects of EDTA-fluoride with EDTA-chloride, as an increase in ionic strength alone may be inhibitory (FOULKES et al. 1983a)

than phosphoseryl-casein. In particular, PPT-2 was 1000 times more active in dephosphorylating phosphotyrosyl-casein.

Since EDTA-fluoride failed to inhibit these enzymes, we examined other putative inhibitors, and found that 30 m*M* pyrophosphate inhibited all three activities by greater than 95%. We also noted that 1 m*M* Mn^{2+}, but not Mg^{2+}, significantly inhibited these activities. This suggested that the apparent preference of certain tyrosyl-protein kinases for MnATP over MgATP (Hunter and Sefton 1981) could be mediated by an inhibition of phosphatase activity, rather than a change in the kinase activity per se.

Recently, Horlein et al. (1982) also have separated multiple phosphotyrosyl-protein phosphatases, using extracts from Ehrlich ascites tumor cells. The principal substrate in these studies was a chemically modified glycogen phosphorylase, phosphorylated in vitro by the EGF receptor kinase from A431 cells. When cell lysates were fractionated on DEAE-Sephadex, 60% of the phosphotyrosyl-protein phosphatase was found in the flow-through fraction. The activity which bound to the column separated into two peaks. After elution, the major activity (30%–35%) was further fractionated on Zn^{2+}-agarose chromatography and again separated into two peaks. The major activity at this step had an M_r of 40000 on gel filtration and was essentially free of phosphoseryl-glycogen phosphatase activity. Micromolar levels of Zn^{2+} were found to inhibit the $M_r = 40000$ enzyme, but the degree of inhibition was dependent on the substrate employed. The dephosphorylation of phosphotyrosyl-membrane proteins from A431 cells was inhibited 50% by 5 μ*M* Zn^{2+}, while 33 μ*M* Zn^{2+} was required to inhibit the dephosphorylation of phosphotyrosyl-phosphorylase by 50%.

In our studies using the chicken brain extract (Foulkes et al. 1983a), approximately 10% of the phosphotyrosyl-casein phosphatase activity fractionated with the phosphoseryl-protein phosphatases $PP2A_1$ and $PP2A_2$, and showed optimal activity in the presence of Mn^{2+}. Chicken brain $PP2A_1$ and $PP2A_2$ migrated on gel filtration with apparent M_r of 250000 and 110000 respectively[1]. However, when these enzymes were treated with ethanol to release the catalytic subunit ($M_r = 35000$) (Lee et al. 1980), both phosphotyrosyl- and phosphoseryl-protein phosphatase activities were again found to co-migrate, but now with an M_r of 35000. Furthermore, the relative phosphotyrosyl- and phosphoseryl-protein phosphatase activities of $PP2A_1$, $PP2A_2$, and the $M_r = 35000$ catalytic subunit were identical. These observations strongly suggested that the type 2A enzymes have an inherent phosphotyrosyl-protein phosphatase activity. Data supporting this have also been obtained by Li and co-workers (Li 1982; Chernoff et al 1983a). In these studies essentially homogeneous preparations of bovine cardiac $PP2A_2$[2] were found to dephosphorylate both phosphotyrosyl-IgG and

1 The molecular weight of $PP2A_2$ has been variously reported between 80000 and 150000 (Tamura et al. 1980; Crouch and Safer 1980; Cohen et al. 1981; Foulkes et al. 1983a; Ingebritsen et al. 1983). Whether this reflects tissue variation, the presence of tetrameric and dimeric forms, or both, remains to be determined

2 Dr. Li's laboratory has previously designated $PP2A_2$ and PP2C as type 1 and type 2 respectively. This reflects differences in nomenclature only, and not enzyme type. For the sake of clarity, the former nomenclature has been retained throughout the text

phosphotyrosyl-casein. This activity was inhibited by EDTA and stimulated by Mn^{2+}, properties characteristic of the type 2A enzymes using phosphoseryl-proteins as substrates. Furthermore, whether $PP2A_2$ existed as the $M_r = 95000$ form or the $M_r = 35000$ catalytic subunit, both the phosphoseryl- and the phosphotyrosyl-protein phosphatase activities comigrated on native acrylamide gel electrophoresis. However, these two activities demonstrated somewhat different properties under certain conditions. For example, heat treatment at 40 °C for 60 min almost completely inactivated the phosphotyrosyl-protein phosphatase activity, while having little effect on the phosphoseryl-protein phosphatase activity[1].

The molecular weight of PPT-2 ($M_r = 43000$) is similar to that reported for the phosphoseryl-protein phosphatase, PP2C (LI et al. 1978; LI 1981, 1982). In addition, LI has reported that highly purified PP2C is active against phosphotyrosyl-IgG (LI et al. 1981; LI 1982; CHERNOFF et al., to be published). However, PPT-2 and PP2C appear to be distinct enzymes. PPT-2 shows optimal activity in the presence of EDTA (FOULKES et al. 1983a), while PP2C requires metal ions for either its phosphotyrosyl- or its phosphoseryl-protein phosphatase activity. In our studies on chicken brain phosphatases (FOULKES et al. 1983a) we failed to observe significant levels of PP2C and, therefore, the importance of this enzyme as a phosphotyrosyl-protein phosphatase could not be assessed. In LI's study using bovine cardiac extracts, where PP2C is a major phosphoseryl-protein phosphatase activity, PP2C contained only 0.5% of the phosphotyrosyl-IgG phosphatase activity and exhibited even less activity toward phosphotyrosyl-casein (CHERNOFF et al., to be published).

Taken together, these studies indicate that the type 2A and the type 2C phosphoseryl-protein phosphatases can act as phosphotyrosyl-protein phosphatases in vitro using artificial substrates. However, the quantitative significance of the phosphotyrosyl-protein phosphatase activity of these enzymes remains to be determined.

The ability of type 1 phosphoseryl-protein phosphatases to act as phosphotyrosyl-protein phosphatases has also been investigated, but the data are more preliminary. In our studies of chicken brain phosphatases (FOULKES et al. 1983a), no significant amount of phosphotyrosyl-protein phosphatase activity was found in the fractions containing PP1. However, in this tissue, PP1 represents a relatively minor phosphoseryl-protein phosphatase activity. In rat skeletal muscle, which contains a much higher activity of PP1 than does chicken brain, addition of inhibitor-2, which specifically inhibits type 1 phosphatases, had no effect on the dephosphorylation of phosphotyrosyl-IgG (FOULKES et al. 1981). LI's group has also reported that the MgATP-dependent PP1 had no activity toward either phosphotyrosyl IgG or phosphotyrosyl-casein (CHERNOFF et al. 1983). In the above studies comparatively small amounts of PP1 were employed and further studies are required to determine whether the type 1 phosphatases are completely devoid of phosphotyrosyl-protein phosphatase activity.

1 A similar effect has been observed with PP1, whereby pretreatment of the enzyme with EDTA inhibits the dephosphorylation of some substrates but not others (NIMMO and COHEN 1978a, b)

2.2.3 The Relation of Phosphotyrosyl-Protein Phosphatases to Acid and Alkaline Phosphatases

LEIS and KAPLAN (1982) isolated an acid phosphatase from a human tumor astrocytoma, which copurified with a phosphotyrosyl-histone phosphatase activity. Although these authors concluded that the acid and protein phosphatase activities were due to the same enzyme, certain observations suggest they may be due to separate molecular entities; in particular *p*-nitrophenylphosphate, used as the substrate to determine the acid phosphatase activity, did not significantly inhibit the dephosphorylation of phosphotyrosyl-histones. Based on the concentrations of these two substrates in the assay and their estimated K_m values, 85% inhibition should have been observed due to the presence of a competitive substrate.

SWARUP et al. (1981) first demonstrated that a variety of alkaline phosphatases, including preparations from calf intestine and *Escherichia coli,* dephosphorylated phosphotyrosyl-histones at 5–10 times the rate observed with phosphoseryl-histones as substrate. These enzymes were also active against the phosphotyrosyl-proteins in membranes from A431 cells. Although the alkaline phosphatase and phosphotyrosyl-protein phosphatase activities were shown to comigrate on high-performance liquid chromatography, the pH optimum for the latter activity was very broad, with more than 70% of the activity being expressed over the range pH 5.5–9.0.

In TCRC-2 cells (a cell line derived from HeLa), 70% of the phosphotyrosyl-histone phosphatase activity was reported to be associated with a particulate fraction (SWARUP et al. 1982b). This activity was solubilized with detergent and subjected to fractionation on wheat germ lectin and histone-Sepharose chromatography. On both columns, the phosphotyrosyl-protein phosphatase activity eluted with an alkaline phosphatase activity. One m*M* *p*-nitrophenylphosphate inhibited the dephosphorylation of phosphotyrosyl-histones by 90%, as expected for a competitive substrate, while 100 μ*M* Zn^{2+} had relatively little effect. Ten μ*M* orthovanadate inhibited both alkaline and phosphotyrosyl-protein phosphatase activities by more than 80%. Based on such observations, SWARUP et al. (1981; 1982a, b) suggested that dephosphorylation of phosphotyrosyl-proteins represents the physiological role for alkaline phosphatases. However, this appears unlikely for several reasons.

Alkaline phosphatases are Zn^{2+} metalloproteins, which also require Mg^{2+} for optimal activity (FERNLEY 1971). EDTA is a potent inhibitor of these enzymes using either *p*-nitrophenylphosphate (FERNLEY 1971) or phosphotyrosyl-proteins (SWARUP et al. 1981; FOULKES et al. 1983a) as the substrate. However, as discussed above, EDTA appears to have little effect on the major phosphotyrosyl-protein phosphatase activities in a variety of tissue and cell extracts (BRAUTIGAN et al. 1981; FOULKES et al. 1981; GALLIS et al. 1981; HORLEIN et al. 1982; LEIS and KAPLAN 1982; FOULKES et al. 1983a). The fact that both alkaline and acid phosphatases from a variety of sources appear to act upon phosphotyrosyl-proteins in vitro (SWARUP et al. 1981; LEIS and KAPLAN 1982) probably reflects the structural similarity of *p*-nitrophenylphosphate and phosphotyrosine, rather than an important physiological activity. Of the five phosphotyrosyl-protein phosphatases we isolated from chicken brain extracts, all showed considerably more activity[1] than bovine intestinal alkaline phosphatase (FOULKES et al. 1983a); in particular, PPT-2 dephosphorylated phosphotyrosyl-casein 3000 times more rapidly than observed using the alkaline phosphatase preparation. Teleologically speaking, since phosphotyrosine is scarce in normal cells (HUNTER and SEFTON 1981), it would appear uneconomical for certain tissues to contain close to millimolar levels of alkaline phosphatase (FERNLEY 1971).

LI has suggested that the type 2A phosphatases possess an inherent alkaline phosphatase activity (LI 1981, 1982) which is distinct from the classical alkaline phosphatase of membranes (FERNLEY 1971). However, as discussed above, the phosphotyrosyl-protein phosphatase activity of the type 2A enzymes appears to represent only a minor activity, using a variety of substrates and extracts. In conclusion, alkaline phosphatases appear to be distinct from the major phosphotyrosyl-protein phosphatase activities.

3 Phosphotyrosine Phosphatases

Free phosphotyrosine is present at millimolar levels in *Drosophila* larvae and disappears rapidly before pupation (MITCHELL and LUNAN 1964; FUKAMI and LIPMANN 1982). However, the existence of free phosphotyrosine in vertebrate cells has not yet been reported. This compound appears to act as a tyrosine storage form in *Drosophila,* but other roles have not been excluded. FUKAMI and LIPMANN (1982) recently isolated an M_r of 170000 phosphatase from *Drosophila* larvae using free L-phosphotyrosine as a substrate. This phosphatase was particularly interesting in that it appears to be able to catalyze a phosphate exchange between tyrosine phosphate and tyrosine. A number of seryl-protein kinases also have been shown to act reversibly, catalyzing dephosphorylation of proteins coupled to the synthesis of ATP from ADP (SHIZUTA et al. 1975; 1977). FUKAMI and LIPMANN (1983) recently reported that the free energy of hydrolysis of phosphotyrosine bound to IgG was similar to that of the γ phosphate in ATP, while the heat of hydrolysis of free phosphotyrosine was three- to fourfold lower. Furthermore, they showed that under the appropriate assay conditions and in the presence of ADP, the tyrosyl-protein kinase of Rous sarcoma virus could act in a readily reversible manner to dephosphorylate phosphotyrosyl-IgG coupled to the synthesis of ATP. It will be of interest to test other phosphotyrosyl-proteins for the presence of a high-energy phosphate bond.

1 These experiments were carried out by normalizing the alkaline phosphatase activity of each fraction and then measuring their relative activities toward phosphotyrosyl-casein in the standard assay

4 Conclusions

The main limitation of all of the studies described above is the use of artificial phosphotyrosyl-proteins as substrates. To date, this approach has been necessary, since physiological substrates (a) are not clearly identified, (b) are poorly phosphorylated in vitro, or (c) occur in such limited amounts as to preclude their use in routine assays. Although a few studies have employed A431 cell membranes containing the autophosphorylated EGF receptor, the function of this phosphoprotein, if any, remains to be determined. It is of interest to remember that while many protein kinases can undergo autophosphorylation in vitro, the physiological significance of these reactions has yet to be demonstrated in vivo.

The artificial nature of these substrates clearly places a restriction on the interpretation of the data, as such studies may reveal activities which have no relevance in vivo. Subject to this limitation, we may draw the following conclusions.

Multiple phosphotyrosyl-protein phosphatases have been indentified in several cell types. In our studies using avian brain extracts (FOULKES et al. 1983a) and those of HORLEIN et al. (1982) using Ehrlich ascites tumor cells, three major and two minor phosphotyrosyl-protein phosphatases were separated chromatographically. Similar results have also been obtained by Li's laboratory (CHERNOFF and LI 1983) using bovine cardiac tissue, where three major activities were separated by chromatography on DEAE-cellulose. Although each study employed different substrates and different cell extracts, a major activity, accounting for 35–56% of the total, was found to elute at a similar position on ion-exchange chromatography. Each preparation showed optimal activity in the presence of EDTA and had an M_r of 40000–60000 on gel filtration. If further purification reveals that three independent studies have isolated a single enzyme, this would clearly be suggestive of an important role for this protein.

The major phosphotyrosyl-protein phosphatase activities are independent of metal ions and are insensitive to EDTA-fluoride. Furthermore, these enzymes can be separated, by chromatography, from all the protein phosphatases which have been described previously. These findings clearly indicate that the major phosphotyrosyl-protein phosphatase activities are distinct from phosphoseryl-phosphothreonyl-protein phosphatases.

The reported sensitivity of phosphotyrosyl-protein phosphatases to Zn^{2+} appears to vary widely, but this probably reflects the existence of multiple activities in different cell types and the use of a variety of substrates.

The relevance of alkaline phosphatases as phosphotyrosyl-protein phosphatases in vivo cannot be resolved until more physiological substrates are employed. However, they are unlikely to be important for a number of reasons; notably, unlike phosphotyrosyl-protein phosphatases, alkaline phosphatases are almost completely inhibited by EDTA. It will also be important to determine what fraction of the total phosphotyrosyl-protein phosphatase activity is membrane bound, a question which has received comparatively little attention. Regardless of the role of alkaline phosphatases in vivo, these enzymes should still prove useful to dephosphorylate phosphotyrosyl-proteins in vitro and to determine the functional significance of this modification.

Owing to the lack of data using physiological substrates, it is difficult to assess the importance of the observation that the type 2A and type 2C phosphoseryl-protein phosphatases appear to contain an inherent phosphotyrosyl-protein phosphatase activity. These enzymes appear to represent only minor activities, using a variety of substrates and different cell extracts. However, these results demonstrate that unlike protein kinases, which show a clear specificity for the amino acid phosphate acceptor, protein phosphatases show less specificity and can act on a variety of residues. It is also interesting to recall that all viral tyrosyl-protein kinases are themselves phosphorylated on serine residues (HUNTER and SEFTON 1981). Although the function of phosphoserine in these proteins remains to be determined, such a modification introduces the very interesting possibility that phosphoseryl-protein phosphatases could regulate the levels of phosphotyrosine in cells, by modulating the activity of the tyrosyl-protein kinase.

One of the major problems relating to the role of tyrosine phosphorylation in cell growth and transformation is that the proteins which are phosphorylated on tyrosine in vivo appear to be phosphorylated to only a very low extent (RADKE et al. 1980; ERIKSON et al. 1981; HUNTER and SEFTON 1981; SEFTON et al. 1981; COOPER et al. 1983). A number of proteins are known to undergo a very rapid phosphorylation/dephosphorylation within seconds of breaking open the cells (MANNING et al. 1980). It was, therefore, possible that the degree of phosphorylation of phosphotyrosyl-proteins in vivo was considerably higher than estimated, as specific precautions to stop dephosphorylation were seldom taken. However, studies in our laboratory tend to support the low levels of phosphorylation of phosphotyrosyl-proteins in vivo. Even by homogenizing cells at −20 °C in the presence of a variety of phosphatase inhibitors we have failed to observe any significant increase in tyrosine phosphorylation (FOULKES JG and BALTIMORE D, unpublished observations).

Finally, it is interesting to speculate on the existence of phosphotyrosyl-protein phosphatase inhibitors, analogous to those described for the phosphoseryl-protein phosphatases (FOULKES and COHEN 1980; LEE et al. 1980; TAMURA and TSUIKI 1980; COHEN et al. 1981; FOULKES et al. 1982; LI 1982). If, as argued previously, transformation represents a tyrosyl-protein kinase/phosphatase equilibrium process, then an increase in phosphotyrosine could be brought about by the introduction (or activation) of a gene encoding a phosphotyrosyl-protein phosphatase inhibitor. While such proteins have not yet been described, cells transformed by avian erythroblastosis virus show a 50% increase in the amount of phosphotyrosine in proteins, even though the gene products of this virus do not appear to be associated with a tyrosyl-protein kinase activity (HUNTER and SEFTON 1981).

As indicated by the data obtained with *ts* transforming proteins, reversion to the normal cell phenotype is actually mediated by phosphotyrosyl-protein phosphatases. This suggests the intriguing possibility that sufficient activation of these enzymes could reverse the transformation process even in the presence of an active tyrosyl-protein kinase. It is also evident from these findings that phosphotyrosyl-protein phosphatases have the potential of regulating hormone action, cell growth, and differentiation. Clearly then, isolation, characterization, and an insight into the regulation of these enzymes is essential for our understanding of these processes.

Acknowledgments. I would like to thank my colleagues for their helpful suggestions in writing this review and for making their unpublished data freely available. I would also like to thank Ms. GINGER PIERCE for the typing of this manuscript.

References

Ambros V, Baltimore D (1978) Protein is linked to the 5′ end of poliovirus RNA by a phosphodiester linkage to tyrosine. J Biol Chem 253:5263–5266

Antoniw JF, Cohen P (1976) Separation of two phosphorylase kinase phosphatases from rabbit skeletal muscle. Eur J Biochem 68:45–54

Avruch J, Nemenoff A, Blackshear PJ, Pierce MW, Osathanondh R (1982) Insulin-stimulated tyrosine phosphorylation of the insulin receptor in detergent extracts of human placental membranes. J Biol Chem 257:15162–15166

Beemon K (1981) Transforming proteins of some feline and avian sarcoma viruses are related structurally and functionally. Cell 24:145–153

Bishop MJ (1982) Oncogenes. Sci Am 246:80–93

Bishop JM, Varmus H (1982) Functions and origins of retroviral transforming genes. In: Weiss R, Teich N, Varmus H, Coffin J (eds) RNA tumor viruses, 2nd edn. Cold Spring Harbor Laboratory, New York, pp 999–1108

Brautigan DL, Bornstein P, Gallis B (1981) Phosphotyrosyl-protein phosphatases. J Biol Chem 256:6519–6522

Buhrow SA, Cohen S, Staros JV (1982) Affinity labeling of the protein kinase associated with epidermal growth factor receptor in membrane vesicles from A431 cells. J Biol Chem 257:4019–4022

Carpenter G, King L, Cohen S (1979) Rapid enhancement of protein phosphorylation in A-431 cell membrane preparations by epidermal growth factor. J Biol Chem 254:4884–4891

Chernoff J, Li HC (1983) Multiple forms of phosphotyrosyl-protein phosphatase from bovine heart and brain. Fed Am Soc Exp Biol (in press)

Chernoff J, Li HC, Cheng YS, Chen LB (1983) Characterisation of a phosphotyrosyl-protein phosphatase activity associated with a phosphoseryl-protein phosphatase of $M_r = 95000$ from bovine heart. J Biol Chem (in press)

Chernoff J, Tabarini D, Li HC, Cheng YSE, Chen LB (to be published) Dephosphorylation of phosphotyrosyl-protein by type II phosphoprotein phosphatase.

Cohen P (1978) The role of the cyclic AMP dependent protein kinase in the regulation of glycogen metabolism in mammalian skeletal muscle. Curr Top Cell Regul 14:118–196

Cohen P (ed) (1980) Recently discovered systems of enzyme regulation by reversible phosphorylation. In: Molecular aspects of cellular regulation, vol 1. Elsevier/North-Holland, New York

Cohen P (1982) The role of protein phosphorylation in neural and hormonal control of cellular activity. Nature 296:613–619

Cohen P, Foulkes JG, Goris J, Hemmings BA, Ingebritsen TS, Stewart AA, Strada ST (1981) Classification of protein phosphatases involved in cellular regulation. In: Holzer H (ed) Metabolic interconversion of enzymes 1980. Springer, Berlin Heidelberg New York

Collett MS, Erikson E, Purchio AF, Brugge JS, Erikson RL (1979) A normal cell protein similar in structure and function to the avian sarcoma virus transforming gene product. Proc Natl Acad Sci USA 76:3159–3163

Cooper JA, Hunter T (1981) Changes in protein phosphorylation in Rous sarcoma virus transformed chicken embryo cells. Mol Cell Biol 1:165–178

Cooper JA, Hunter T (1983) Identification and characterisation of cellular targets for tyrosine protein kinases. J Biol Chem 258:1108–1115
Cooper JA, Bowen-Pope DF, Raines E, Ross R, Hunter T (1982) Similar effects of platelet derived growth factor and epidermal growth factor on phosphorylation of tyrosine in cellular proteins. Cell 31:263–273
Cooper JA, Reiss NA, Schwartz RJ, Hunter T (1983) Three glycolytic enzymes are phosphorylated at tyrosine in cells transformed by Rous sarcoma virus. Nature 302:218–223
Crouch D, Safer B (1980) Purification and properties of elF-2 phosphatase. J Biol Chem 255:7918–7924
Curnow RT, Larner J (1979) Hormonal and metabolic control of phosphoprotein phosphatase. In: Litwack G (ed) Biochemical actions of hormones, vol 6. Academic, New York, pp 77–119
Duesberg PM, Bister K (1981) Transforming genes of retroviruses. Cancer 1:111–136
Eckhart W, Hutchinson MA, Hunter T (1979) An activity phosphorylating tyrosine in polyoma T antigen immunoprecipitates. Cell 18:925–933
Ek B, Heldin CH (1982) Characterization of a tyrosine specific kinase activity in human fibroblast membranes stimulated by platelet-derived growth factor. J Biol Chem 257:10486–10492
Ek B, Westermark B, Wasteson A, Heldin CH (1982) Stimulation of tyrosine-specific phosphorylation by platelet-derived growth factor. Nature 295:419–420
Erikson RL, Purchio AF, Erikson E, Collett MS, Brugge JS (1980) Molecular events in cells transformed by Rous sarcoma virus. J Cell Biol 87:319–325
Erikson E, Cook R, Miller GJ, Erikson RL (1981) The same normal cell protein is phosphorylated after transformation by avian sarcoma viruses with unrelated transforming genes. Mol Cell Biol 1:43–50
Eva A, Robbins KC, Andersen PR, Srinivasan A, Tronick SR, Reddy EP, Ellmore NW, Galen AT, Lautenberger JA, Papas TS, Westin EH, Wong-Staal F, Gallo RC, Aaronson SA (1982) Cellular genes analogous to retroviral *onc* genes are transcribed in human tumor cells. Nature 295:116–119
Fernley HN (1971) Mammalian alkaline phosphatases. In: Boyer PD (ed), The enzymes, vol 4, 3rd edn, Academic, New York, pp 417–447
Foulkes JG, Cohen P (1979) Hormonal control of glycogen metabolism. Eur J Biochem 97:251–256
Foulkes JG, Cohen P (1980) Regulation of glycogen metabolism. Eur J Biochem 105:195–203
Foulkes JG, Maller JL (1982) In vivo actions of protein phosphatase inhibitor-2 in *Xenopus* oocytes. FEBS Lett 150:155–160
Foulkes JG, Howard RF, Ziemiecki A (1981) Detection of a novel mammalian protein phosphatase with activity for phosphotyrosine. FEBS Lett 130:197–200
Foulkes JG, Cohen P, Strada S, Everson WV, Jefferson LS (1982) Antagonistic effects of insulin and β-adrenergic agonists on the activity of protein phosphatase inhibitor-1 in skeletal muscle of the perfused rat hemicorpus. J Biol Chem 257:12493–12496
Foulkes JG, Erikson E, Erikson RL (1983a) Separation of multiple phosphotyrosyl-protein phosphatases from chicken brain. J Biol Chem 258:431–438
Foulkes JG, Ernst V, Levin DH (1983b) Separation and identification of type 1 and type 2 protein phosphatases from rabbit reticulocyte lysates. J Biol Chem 258:1439–1443
Friis RR, Jockusch BM, Boschek CB, Ziemiecki A, Rusamen H, Bauer H (1979) Transformation-defective, temperature sensitive mutants of Rous sarcoma virus have a reversibly defective src-gene product. Cold Spring Harbor Symp Quant Biol 44:1007–1012
Fukami Y, Lipmann F (1982) Purification of a specific reversible tyrosine-*O*-phosphate phosphatase. Proc Natl Acad Sci USA 79:4275–4279
Fukami Y, Lipmann F (1983) Reversal of Rous sarcoma-specific immunoglobulin phosphorylation on tyrosine (ADP as phosphate acceptor) catalysed by the *src* gene kinase. Proc Natl Acad Sci USA 80:1872–1876
Gallis B, Bornstein P, Brautigan DL (1981) Tyrosylprotein kinase and phosphatase activities in membrane vesicles from normal and Rous sarcoma virus-transformed rat cells. Proc Natl Acad Sci USA 78:6689–6693
Goyette M, Petropoulos CJ, Shank PR, Fausto N (1983) Expression of a cellular oncogene during liver regeneration. Science 219:510–512

Green GA, Chenoweth M, Dunn A (1980) Adrenal glucocorticoid permissive regulation of muscle glycogenolysis. Proc Natl Acad Sci USA 77:5711–5715
Hanafusa H (1977) Cell transformation by RNA tumor viruses. Compr Virol 10:401–481
Hemmings BA, Yellowlees D, Kernohan JC, Cohen P (1981) Purification of glycogen synthase kinase-3 from rabbit skeletal muscle. Eur J Biochem 119:443–451
Hemmings BA, Resink T, Cohen P (1982) Reconstruction of a MgATP dependent phosphatase and its activation through a phosphorylation mechanism. FEBS Lett 150:319–324
Hoffman-Falk H, Einat P, Shilo B, Hoffman FM (1983) *Drosophila melanogaster* DNA clones homologous to vertebrate oncogenes. Cell 32:589–598
Hörlein D, Gallis B, Brautigan DL, Bornstein P (1982) Partial purification and characterization of phosphotyrosyl-protein phosphatase from Ehrlich ascites tumor cells. Biochemistry 21:5577–5584
Hunter T, Sefton BM (1981) Protein kinases and viral transformation. In: Cohen P, Van Heyningen S (eds) Molecular aspects of cellular regulation, vol 2. Elsevier-North Holland, New York, pp 337–370
Hunter T, Sefton BM, Beemon K (1979) Studies on the structure and function of the avian sarcoma virus transforming gene product. Cold Spring Harbor Symp Quant Biol 44:931–941
Ingebritsen TS, Foulkes JG, Cohen P (1980) The broad specificity protein phosphatase from mammalian liver. FEBS Lett 119:9–15
Ingebritsen TS, Foulkes JG, Cohen P (1983a) Classification of protein phosphatases involved in the regulation of glycogen metabolism. Eur J Biochem 132:263–274
Ingebritsen TS, Stewart AA, Cohen P (1983b) Measurement of type 1 and type 2 protein phosphatases in extracts of mammalian tissues. Eur J Biochem 132:297–307
Kasuga M, Zick Y, Blithe DL, Crettaz M, Kahn CR (1982a) Insulin stimulates tyrosine phosphorylation of the insulin receptor in a cell-free system. Nature 228:667–669
Kasuga M, Zick Y, Blith DL, Karlsson FA, Haring HU, Kahn CR (1982b) Insulin stimulation of phosphorylation of the β subunit of the insulin receptor. J Biol Chem 257:9891–9894
Krebs EG, Beavo JA (1979) Phosphorylation-dephosphorylation of enzymes. Annu Rev Biochem 48:923–959
Lee EYC, Silberman SR, Ganapathi MK, Petrovic S, Paris H (1980) The phosphoprotein phosphatases. Adv Cyclic Nucleotide Res 13:95–131
Lee EYC, Silberman SR, Ganapathi MK, Paris H, Petrovic S (1981) Properties of rabbit skeletal muscle protein phosphatases. In: Rosen OM, Krebs EG (eds) Cold Spring Harbor conferences on cell proliferation, vol 8. Cold Spring Harbor, New York, pp 425–439
Lee W, Bister K, Moscovici C, Duesberg P (1981) Temperature-sensitive mutants of Fujinami sarcoma virus. J Virol 38:1064–1076
Leis JF, Kaplan NO (1982) An acid phosphatase in the plasma membranes of human astrocytoma showing marked specificity toward phosphotyrosine protein. Proc Natl Acad Sci USA 79:6507–6511
Li HC (1981) Purification and properties of cardiac muscle phosphoprotein phosphatase and alkaline phosphatase isoenzymes. In: Rosen OM, Krebs EG (eds) Cold Spring Harbor conferences on cell proliferation, vol 8. Cold Spring Harbor, New York, pp 441–457
Li HC (1982) Phosphoprotein phosphatases. Curr Top Cell Regul 21:129–173
Li HC, Hsiao KJ, Chan WWS (1978) Purification and properties of phosphoprotein phosphatases from canine heart. Eur J Biochem 84:215–225
Li HC, Tabarini D, Cheng YS, Chen LB (1981) Dephosphorylation of pp60^{src}-protein kinase phosphorylated immunoglobulin by type II phosphoprotein phosphatase. Fed Am Soc Exp Biol 40:1539 (Abstr 5)
Linial M, Blair D (1982) Genetics of retroviruses. In: Weiss R, Teich N, Varmus H, Coffin J (eds) RNA tumor viruses, 2nd edn. Cold Spring Harbor Laboratory, New York, pp 649–784
Manning DR, DiSalvo J, Stull JT (1980) Protein phosphorylation: quantitative analysis in vivo and in intact cell systems. Mol Cell Endocrinol 19:1–19
Mitchell HK, Lunan KD (1964) Tyrosine-*O*-phosphate in Drosophila. Arch Biochem Biophys 106:219–222

Muller R, Slamon DJ, Tremblay JM, Cline MJ, Verma IM (1982) Differential expression of cellular oncogenes during pre- and post-natal development of the mouse. Nature 299:640–644

Nimmo GA, Cohen P (1978a) Regulation of glycogen metabolism. Eur J Biochem 87:341–351

Nimmo GA, Cohen P (1978b) Regulation of glycogen metabolism. Eur J Biochem 87:353–365

Nimmo HG, Cohen P (1977) Hormonal control of protein phosphorylation. Adv Cyclic Nuceotide Res 8:146–260

Pato M, Adelstein RS (1980) Dephosphorylation of the 20000 dalton light chain of myosin by two different phosphatases from smooth muscle. J Biol Chem 255:6535–6538

Pato M, Adelstein RS, Crouch DB, Safer B, Ingebritsen TS, Cohen P (1983) Classification of two homogeneous myosin light chain phosphatases and a homogeneous protein phosphatase from reticulocytes. Eur J Biochem 132:283–287

Radke K, Martin GS (1979) Transformation by Rous sarcoma virus: effects of src-gene expression on the synthesis and phosphorylation of cellular polypeptides. Cold Spring Harbor Symp Quant Biol 44:975–982

Radke K, Gilmore T, Martin GS (1980) Transformation by Rous sarcoma virus: a cellular substrate for transformation-specific protein phosphorylation contains phosphotyrosine. Cell 21:821–828

Resink T, Hemmings BA, Lim Tung HY, Cohen P (1983) Characterisation of a reconstituted MgATP dependent protein phosphatase. Eur J Biochem 133:455–461

Reynolds FH, Todaro GJ, Fryling C, Stephenson JR (1981) Human transforming growth factors induce tyrosine phosphorylation of EGF receptor. Nature 292:259–262

Rothberg PG, Harris TJR, Nomoto A, Wimmer E (1978) O^4-(5′-uridylyl)tyrosine is the bond between the genome-linked protein and the RNA of poliovirus. Proc Natl Acad Sci USA 75:4868–4872

Schartl M, Barnekow A (1982) Expression in eukaryotes of a tyrosine kinase which is reactive with $pp60^{v\text{-}src}$ antibodies. Differentiation 23:109–114

Sefton BM, Hunter T, Beeman K, Eckhart W (1980) Evidence that the phosphorylation of tyrosine is essential for cellular transformation by Rous sarcoma virus. Cell 20:807–816

Sefton BM, Hunter T, Ball EH, Singer SJ (1981) Vinculin: a cytoskeletal target of the transforming protein of Rous sarcoma virus. Cell 24:165–174

Shapiro BM, Stadtman ER (1968) 5′Adenylyl-*O*-tyrosine. J Biol Chem 243:3769–3771

Shizuta Y, Beavo JA, Bechtel PJ, Hoffmann F, Krebs EG (1975) Reversibility of adenosine 3′-5′monophosphate dependent protein kinase reactions. J Biol Chem 250:6891–6896

Shizuta Y, Khandelwal RJ, Maller JL, Van Den Heede JR, Krebs EG (1977) Reversibility of phosphorylase kinase reaction. J Biol Chem 252:3408–3413

Spector DH, Varmus HE, Bishop JM (1978) Nucleotide sequences related to the transforming gene of avian sarcoma virus are present in DNA of uninfected vertebrates. Proc Natl Acad Sci USA 75:4102–4106

Stewart AA, Hemmings BA, Cohen P, Goris J, Merlevede W (1981) The MgATP-dependent protein phosphatase and protein phosphatase-1 have identical substrate specificities. Eur J Biochem 115:197–205

Stewart AA, Ingebritsen TS, Manalan A, Klee CB, Cohen P (1982) Discovery of a Ca^{++} and calmodulin-dependent protein phosphatase. FEBS Lett 137:80–84

Stewart AA, Ingebritsen TS, Cohen P (1983) Purification and properties of a calcium ion and calmodulin dependent protein phosphatase 2B from rabbit skeletal muscle. Eur J Biochem 132:289–295

Swarup G, Cohen S, Garbers DL (1981) Selective dephosphorylation of proteins containing phosphotyrosine by alkaline phosphatases. J Biol Chem 256:8197–8201

Swarup G, Cohen S, Garbers DL (1982a) Inhibition of phosphotyrosyl-protein phosphatase activity by vanadate. Biochem Biophys Res Commun 107:1104–1109

Swarup G, Speeg KV, Cohen S, Garbers DL (1982b) Phosphotyrosyl-protein phosphatase of TCRC-2 cells. J Biol Chem 257:7298–7301

Taborsky G (1974) Phosphoproteins. Adv Protein Chem 28:1–210

Tamura S, Tsuiki S (1980) Purification and subunit structure of rat-liver phosphoprotein phosphatase $M_r = 260000$. Eur J Biochem 111:217–224

Tamura S, Kikuchi H, Kikuchi K, Hiraga A, Tsuiki S (1980) Purification and subunit structure of rat liver phosphoprotein phosphatase II. Eur J Biochem 104:347–355
Ushiro H, Cohen S (1980) Identification of phosphotyrosine as a product of epidermal growth factor-activated protein kinase in A431 cell membranes. J Biol Chem 255:8363–8365
Vandenheede JR, Yang SD, Goris J, Merlevede W (1980) ATPMg dependent protein phosphatase from rabbit skeletal muscle. J Biol Chem 255:11768–11774
Vandenheede JR, Goris J, Yang SD, Camps T, Merlevede W (1981a) Conversion of active protein phosphatase to the ATPMg-dependent form enzyme by inhibitor-2. FEBS Lett 127:1–3
Vandenheede JR, Yang SD, Merlevede W (1981b) Rabbit skeletal muscle protein phosphatase. J Biol Chem 256:5894–5900
Varmus HE (1982) Form and function of retroviral proviruses. Science 216:812–820
Walsh DA, Perkins JP, Krebs EG (1968) An adenosine 2′,5′-monophosphate dependent protein kinase from rabbit skeletal muscle. J Biol Chem 243:3763–3765
Weller M (1979) Protein phosphorylation. In: Lagnado JR (ed) Pion, London
Witte O, Dasgupta A, Baltimore D (1980) Abelson murine leukaemia virus protein is phosphorylated in vitro to form phosphotyrosine. Nature 283:826–831
Yang SD, Vandenheede JR, Goris J, Merlevede W (1980) ATPMg dependent protein phosphatase from rabbit skeletal muscle. J Biol Chem 255:11759–11767
Zick Y, Kasuga M, Kahn RC, Roth J (1983) Characterization of insulin mediated phosphorylation of the insulin receptor in a cell-free system. J Biol Chem 258:75–80
Ziemiecki A, Friis RR (1980) Phosphorylation of $pp60^{src}$ and the cyclohexamide insensitive activation of the $pp60^{src}$ associated kinase activity of transformation-defective temperature-sensitive mutants of Rous sarcoma virus. Virology 106:391–394

MIX
Papier aus verantwortungsvollen Quellen
Paper from responsible sources
FSC® C105338

If you have any concerns about our products,
you can contact us on
ProductSafety@springernature.com

In case Publisher is established outside the EU,
the EU authorized representative is:
Springer Nature Customer Service Center GmbH
Europaplatz 3, 69115 Heidelberg, Germany

Printed by Libri Plureos GmbH
in Hamburg, Germany